Charles Dixon

The nests and eggs of non-indigenous British birds or such species

Charles Dixon

The nests and eggs of non-indigenous British birds or such species

ISBN/EAN: 9783337143107

Printed in Europe, USA, Canada, Australia, Japan

Cover: Foto ©berggeist007 / pixelio.de

More available books at **www.hansebooks.com**

THE NESTS AND EGGS

OF

NON-INDIGENOUS BRITISH BIRDS

OR

Such Species that do not Breed within the
British Archipelago

By CHARLES DIXON

AUTHOR OF

'THE GAME BIRDS AND WILD FOWL OF THE BRITISH ISLANDS,'
'THE MIGRATION OF BIRDS,' 'THE NESTS AND EGGS OF BRITISH BIRDS,'
ETC., ETC., ETC.

WITH COLOURED FRONTISPIECE

LONDON: CHAPMAN AND HALL, Ld.

1894

[*All rights reserved*]

RICHARD CLAY & SONS, LIMITED,
LONDON & BUNGAY.

PREFACE.

THE present work forms the companion volume to *The Nests and Eggs of British Birds*, and renders the subject of British Oology complete, so far as our knowledge now extends. It deals exclusively with the nidification of the birds that do not breed in the British Archipelago, but visit our islands regularly in winter, pass the coasts on passage, or pay them their more or less irregular visits as wanderers from their normal areas of dispersal.

I have spared no pains to make the subject as complete and full of information as possible; but, as the student will eventually discover, there still remains a vast amount of work to be done, which will take years and years of further study to accomplish. Of the birds that breed in civilized areas our information is fairly complete, although even here many details are wanting respecting the habits of birds in the pairing and breeding seasons, the number of broods reared, and the duration of the periods of incubation. When we come to deal with species that spend the summer in the Arctic regions, or dwell permanently in deserts, and in countries little explored by the scientific naturalist, our want of information becomes only too woefully apparent. I am sanguine enough to hope that by pointing out the deficiencies in our knowledge, I may stimulate observation, and thus help in one way to reduce them.

For the purposes of the present work I have examined and compared a vast number of nests and eggs, and in nearly every instance the descriptions have been made from actual specimens, in most cases carefully deduced from large and representative series.

At the end of the volume I have added two Appendices, one containing a list of those species which have a just right to be ranked as British, but whose eggs, nests, and habits during the breeding season are as yet unknown to science; the other a list of doubtful British species. I sincerely trust that the above proportionately somewhat long list of birds whose eggs are unknown will speedily decrease; for in this direction lies some of the most attractive and important work yet to be accomplished in the field of British Oology.

CHARLES DIXON.

CONTENTS.

		PAGE
Nutcracker	*Nucifraga caryocatactes* ...	1
Rose-coloured Starling	*Pastor roseus*	4
American White-winged Crossbill	*Loxia leucoptera*	6
European White-winged Crossbill	„ *bifasciata*	7
Parrot Crossbill	„ *curvirostra pityopsittacus*	8
Pine Grosbeak	„ *enucleator*	9
Scarlet Rose Finch	*Carpodacus erythrinus* ...	11
Canary	*Serinus hortulanus canarius*	13
Serin Finch	„ *hortulanus* ...	14
Brambling	*Fringilla montifringilla* ...	16
Mealy Redpole	*Linota linaria*	18
Greenland Redpole	„ *linaria hornemanni*	20
Lapland Bunting	*Calcarius lapponicus* ...	21
Black-headed Bunting	*Emberiza melanocephala* ...	23
Ortolan Bunting	„ *hortulana* ...	25
Little Bunting	„ *pusilla* ...	27
Shore-Lark	*Otocoris alpestris* ...	29
Calandra Lark	*Melanocorypha calandra* ...	31
White-winged Lark	„ *sibirica* ...	33
Short-toed Lark	*Calandrella brachydactyla*	35
Crested Lark	*Galerita cristata*	37
Alpine Pipit	*Anthus spipoletta* ...	39
Tawny Pipit	„ *campestris* ...	41
Richard's Pipit	„ *richardi* ...	43
Red-throated Pipit	„ *cervinus* ...	45
Wall-Creeper	*Tichodroma muraria* ...	47
Firecrest	*Regulus ignicapillus* ...	49
Waxwing	*Ampelis garrulus* ...	51

CONTENTS.

		PAGE
Lesser Gray Shrike	*Lanius minor*	53
Great Gray Shrike	„ *excubitor*	55
Yellow-browed Willow Wren	*Phylloscopus superciliosus*	57
Orphean Warbler	*Sylvia orphea*	59
Barred Warbler	„ *nisoria*	61
Rufous Warbler	*Aëdon galactodes*	63
Great Reed Warbler	*Acrocephalus turdoides*	65
Aquatic Warbler	„ *aquaticus*	67
Icterine Warbler	*Hypolais hypolais*	69
White's Ground Thrush	*Geocichla varia*	71
Fieldfare	*Turdus pilaris*	73
Redwing	„ *iliacus*	75
Black-throated Ouzel	*Merula atrigularis*	78
Arctic Blue-throated Robin	*Erithacus suecica*	79
Isabelline Wheatear	*Saxicola isabellina*	81
Black-throated Wheatear	„ *stapazina*	83
Desert Wheatear	„ *deserti*	84
Rock Thrush	*Monticola saxatilis*	86
Alpine Accentor	*Accentor alpinus*	88
Black-bellied Dipper	*Cinclus aquaticus melanogaster*	89
Red-breasted Flycatcher	*Muscicapa parva*	90
Purple Martin	*Progne purpurea*	92
Great Spotted Cuckoo	*Coccystes glandarius*	93
Black-billed Cuckoo	*Coccyzus erythrophthalmus*	96
Yellow-billed Cuckoo	„ *americanus*	97
White-bellied Swift	*Cypselus melba*	99
Red-necked Nightjar	*Caprimulgus ruficollis*	101
Egyptian Nightjar	„ *ægyptius*	102
Bee-eater	*Merops apiaster*	103
Roller	*Coracias garrulus*	105
Belted Kingfisher	*Ceryle alcyon*	107
Tengmalm's Owl	*Nyctala tengmalmi*	109
Little Owl	*Athene noctua*	110
Snowy Owl	*Nyctea nyctea*	112
Hawk Owl	*Surnia funera*	114
Scops Owl	*Scops scops*	116
Eagle Owl	*Bubo maximus*	117
Griffon Vulture	*Gyps fulvus*	119

CONTENTS.

		PAGE
Egyptian Vulture	... *Neophron percnopterus*	... 121
White Jer-Falcon	... *Hierofalco candicans*	... 123
Iceland Jer-Falcon	„ *islandus*	... 125
Scandinavian Jer-Falcon	„ *gyrfalco*	... 127
Orange-legged Hobby ...	... *Falco vespertinus* ...	... 129
Lesser Kestrel	„ *cenchris* ...	... 131
Spotted Eagle	... *Aquila nævia* ...	... 133
Black Kite	... *Milvus ater* ...	... 135
Swallow-tailed Kite ...	... *Elanoides furcatus*	... 137
Rough-legged Buzzard ...	... *Archibuteo lagopus*	... 139
Goshawk	... *Astur palumbarius*	... 141
American Goshawk	„ *atricapillus* ...	... 142
Hooper Swan	... *Cygnus musicus* ...	... 144
Bewick's Swan	„ *bewicki* ...	... 145
Lesser Snow Goose ...	... *Chen hyperboreus* ...	... 147
Bean Goose	... *Anser segetum* ...	... 148
Pink-footed Goose	„ *brachyrhynchus*	... 150
White-fronted Goose	„ *albifrons* ...	... 151
Lesser White-fronted Goose ...	„ *albifrons minutus* ...	153
Brent Goose	... *Bernicla brenta* ...	... 154
White-bellied Brent Goose ...	„ *brenta glaucogaster*	... 156
Bernacle Goose	„ *leucopsis* ...	... 157
Red-breasted Goose	„ *ruficollis* ...	... 158
Ruddy Sheldrake	... *Tadorna casarca* ...	... 160
American Wigeon	... *Anas americana* ...	... 161
American Teal	„ *carolinensis* ...	... 163
Blue-winged Teal	„ *discors* ...	... 164
Red-crested Pochard ...	... *Fuligula rufina* ...	... 166
White-eyed Pochard	„ *nyroca* ...	... 167
Scaup	„ *marila* ...	... 169
Harlequin Duck	„ *histrionica* ...	... 171
Long-tailed Duck	„ *glacialis* ...	... 173
Velvet Scoter	„ *fusca* ...	... 175
Surf Scoter	„ *perspicillata* ...	... 176
Buffel-headed Duck	... *Clangula albeola* ...	... 178
Golden-eye	„ *glaucion* ...	... 179
Steller's Eider	... *Somateria stelleri* ...	... 181
King Eider	„ *spectabilis*	... 183

CONTENTS.

		PAGE
Hooded Merganser	*Mergus cucullatus*	185
Smew	„ *albellus*	186
Flamingo	*Phœnicopterus roseus*	188
Glossy Ibis	*Plegadis falcinellus*	191
Spoonbill	*Platalea leucorodia*	193
White Stork	*Ciconia alba*	195
Black Stork	„ *nigra*	197
Little Bittern	*Ardetta minuta*	199
American Bittern	*Botaurus lentiginosus*	200
Night Heron	*Nycticorax griseus*	202
Buff-backed Heron	*Ardea bubulcus*	204
Squacco Heron	„ *comata*	206
Little Egret	„ *garzetta*	207
Great White Egret	„ *alba*	209
Purple Heron	„ *purpurea*	210
Common Crane	*Grus communis*	212
Demoiselle Crane	„ *virgo*	214
Great Bustard	*Otis tarda*	216
Little Bustard	„ *tetrax*	218
Macqueen's Bustard	„ *macqueeni*	219
Cream-coloured Courser	*Cursorius gallicus*	221
Common Pratincole	*Glareola pratincola*	223
Sociable Lapwing	*Vanellus gregarius*	225
Killdeer Plover	*Ægialitis vocifera*	226
Ringed Plover	„ *hiaticula*	227
Little Ringed Plover	„ *minor*	229
Caspian Sand Plover	*Ægialophilus asiaticus*	231
Gray Plover	*Charadrius helveticus*	232
Asiatic Golden Plover	„ *fulvus*	234
American Golden Plover	„ *fulvus americanus*	236
Common Stilt	*Himantopus melanopterus*	237
Common Avocet	*Recurvirostra avocetta*	239
Eskimo Whimbrel	*Numenius borealis*	241
Gray Phalarope	*Phalaropus fulicarius*	243
Bartram's Sandpiper	*Totanus bartrami*	245
Spotted Sandpiper	„ *macularius*	246
Green Sandpiper	„ *ochropus*	248
Yellow-legged Sandpiper	„ *flavipes*	250

CONTENTS. xi

		PAGE
Dusky Redshank	*Totanus fuscus*	252
Black-tailed Godwit	*Limosa melanura*	254
Bar-tailed Godwit	„ *rufa*	256
Red-breasted Snipe	*Ereunetes griseus*	257
Turnstone	*Strepsilas interpres* ...	259
Bonaparte's Sandpiper	*Tringa fuscicollis*	261
Purple Sandpiper	„ *maritima*	262
Broad-billed Sandpiper	„ *platyrhyncha* ...	264
American Pectoral Sandpiper ...	„ *acuminata pectoralis*	266
Little Stint	„ *minuta*	267
American Stint	„ *subminuta minutilla*	270
Temminck's Stint	„ *temmincki*	272
Sanderling	„ *arenaria*	274
Buff-breasted Sandpiper	„ *rufescens*	276
Great Snipe	*Scolopax major*	278
Jack Snipe	„ *gallinula* ...	280
Buffon's Skua	*Stercorarius buffoni* ...	282
Pomatorhine Skua	„ *pomatorhinus*	283
Ivory Gull ... ·	*Pagophila eburnea* ...	285
Iceland Gull	*Larus leucopterus*	287
Glaucous Gull	„ *glaucus*	288
Great Black-headed Gull	„ *ichthyaëtus*	290
Mediterranean Black-headed Gull	„ *melanocephalus* ...	291
Bonaparte's Gull	„ *philadelphia* ...	293
Little Gull	„ *minutus*	295
Sabine's Gull	*Xema sabinii*	297
Noddy Tern	*Anous stolidus*	298
Sooty Tern	*Sterna fuliginosa*	300
Caspian Tern	„ *caspia*	302
Gull-billed Tern	„ *anglica*	304
Whiskered Tern	*Hydrochelidon hybrida* ...	306
Black Tern	„ *nigra* ...	308
White-winged Black Tern	„ *leucoptera* ...	310
Brunnich's Guillemot	*Uria troile brunnichi* ...	311
Little Auk	*Mergulus alle*	313
Great Auk	*Alca impennis*	314
Bulwer's Petrel ...	*Bulweria columbina*	316

CONTENTS.

		PAGE
Wilson's Petrel	*Oceanites wilsoni*	317
Sooty Shearwater	*Puffinus griseus*	319
Dusky Shearwater	,, *obscurus*	320
Great Northern Diver	*Colymbus glacialis*	322
Black-necked Grebe	*Podiceps nigricollis*	324
Sclavonian Grebe	,, *cornutus*	326
Red-necked Grebe	,, *rubricollis*	328
Little Crake	*Crex parva*	330
Eastern Turtle Dove	*Turtur orientalis*	331
Appendix I.		334
Appendix II.		345
Index		361

ADDENDUM.

Page 321.—Since writing the account of the nidification of the Stormy Petrel (*Procellaria pelagica*) in my work on *The Nests and Eggs of British Birds*, I have received information of this species nesting in Tor Bay, which very considerably extends its breeding area in England to the eastward. So far as I can ascertain, Lundy Island was previously its most easterly known limit in England, but Mr. Else, the Curator of the Torquay Museum, informs me that a bird with its egg was obtained from a small hole on the Thatcher Rock in Tor Bay. The bird is now in that museum, and the egg is, or was, in the possession of its finder, a resident of Torquay.

THE NESTS AND EGGS

OF

NON-INDIGENOUS BRITISH BIRDS.

Family CORVIDÆ.   Genus NUCIFRAGA.

NUTCRACKER.

NUCIFRAGA CARYOCATACTES (*Linnæus*).

(British: Rare nomadic autumn migrant.)

Single Brooded. Laying season, March to June, according to latitude.

BREEDING AREA: Palæarctic region. The Nutcracker breeds throughout the conifer forests of Europe and Asia, almost as far north as the limits of forest growth, and from the Atlantic to the Pacific. It breeds in the forests of Scandinavia, south of about lat. 67°, on various islands in the Baltic, and thence across Russia and Siberia to Kamtschatka and the Kurile Islands. Southwards in Europe it is known to breed in East Prussia, in the Black Forest, in the Alps and the Carpathians, and possibly in the Pyrenees and the mountains of South Spain. In Asia, although apparently absent from the Caucasus, it breeds on the various southern and central mountain ranges, but its distribution in the nesting season is by no means clearly defined.

BREEDING HABITS: Little more than thirty years ago the habits of the Nutcracker during the breeding season, together with the bird's nest and eggs, were utterly unknown to British oologists. Consequently the most absurd ideas prevailed, and were perpetuated even so recently as the second edition of Morris's *History of British Birds*, published in 1870, wherein the Nutcracker is actually stated to nest in holes of trees! Professor Newton had the honour of introducing the nest and eggs of the Nutcracker to British naturalists from specimens collected by Pastor Theobald on the island of Bornholm in the Baltic Sea; and an account of the interesting discovery was published in the *Proceedings of the Zoological Society* (1862, p. 207; 1867, p. 162, pl. 15, fig. 2). In the former year, however, another account of the nest and eggs of the Nutcracker was published in the *Ibis* (1862, p. 365), being a translation of Herr Schütt's communication to the *Journal für Ornithologie*. In Europe the Nutcracker appears everywhere to be a very early breeder, commencing to nest long before the snow is off the ground. Its favourite breeding haunts are pine and fir forests, and owing to the density of the trees, and the difficulty of getting about in the deep snow, or half-frozen slush, the nest is by no means an easy one to find. Not only so, the birds do not appear to be in any way gregarious during the breeding season, nesting in isolated pairs, and are remarkably shy and retiring throughout the entire period. It is probable that the Nutcracker pairs for life, although a new nest is made each season. During the days of nest-building and courtship the bird is described as very noisy, but as soon as the eggs are laid it becomes silent and wary in the extreme, flitting about the forests in a most stealthy manner. The site for the nest is generally in a moderate-sized spruce, fir, or pine tree at from fifteen to thirty feet from the ground. The nest is made on one of the nearly horizontal branches close to the trunk, and is a bulky open structure composed of larch, fir, and birch twigs, often cemented with a little mud or clay, and lined with moss, roots, strips of bark, and grass either dry or semi-green. The Nutcracker makes little or no demonstration when disturbed at the nest, slipping away very quietly; the female usually sits remarkably close, remaining brooding over her eggs until she is almost touched by the hand, or the stem of the tree is smartly knocked.

RANGE OF EGG COLOURATION AND MEASUREMENT: The eggs of the Nutcracker are from three to five in number, but three appears to be the average clutch, and occasionally only two. They range from very pale greenish-blue to creamy-white in ground colour, freckled and spotted over the entire surface with olive-brown and underlying markings of pale gray. Very exceptionally a fine hair-like streak of darker brown occurs. The amount of marking varies considerably, and on some specimens most of the spots are on the larger half of the egg; on some varieties the pale gray underlying markings predominate and the surface spots are few and large. Average measurement, 1·3 inch in length, by ·95 inch in breadth. Incubation, performed by the female, is presumed to last from sixteen to eighteen days.

DIAGNOSTIC CHARACTERS: The eggs of the Nutcracker so closely resemble the paler varieties of those of the Magpie that they cannot with absolute certainty be distinguished. Under a microscope the shell of the Nutcracker's egg is of much finer grain than that of the Jackdaw, another species with which it may be confused; it is also more fragile and the spots are never so intense. The site and description of the nest prevents any chance of confusion in the field.

Family STURNIDÆ.   Genus PASTOR.

ROSE-COLOURED STARLING.

PASTOR ROSEUS (*Linnæus*).

(British : Abnormal autumn migrant.)

Single Brooded. Laying season, June.

BREEDING AREA: South-central Palæarctic region. The Rose-coloured Starling is one of those nomadic species that seldom resorts to one particular district for breeding purposes for many seasons in succession, and generally chooses a fresh locality each year, resorting to the nearest spot suitable for the purpose in that area which it may chance to be, and to which it has been attracted by the presence of locusts, its favourite food. Great numbers usually breed in company. The most westerly breeding colony of this species was situated at Villafranca, in the Italian province of Verona. Eastwards it breeds irregularly in Bulgaria, the Dobrudscha, South Russia, the Caucasus, Asia Minor, Turkestan, and South Siberia as far east as Lake Saisan. It may possibly yet be found breeding in Palestine and Northern Persia, both of which countries it traverses on migration to and from its winter quarters in India.

BREEDING HABITS: If the Rose-coloured Starling pairs for life, as is not improbable, seeing that its relation, the Common Starling, does so, and the habits of the two species are much alike, it rarely if ever returns to the old breeding place for two seasons in succession. Like that bird, it is gregarious during the breeding season, nesting in colonies often of enormous dimensions, and apparently breeding in localities where locusts, its favourite food, chance to be abundant. As soon as the young are able to fly the nesting places are deserted. The colonies are established in old ruins, amongst the masonry of castles, in deserted and crumbling earthworks, in the sides of quarries and railway cuttings, in cliffs, or even amongst loose stones and fragments of rock on bare mountain sides. The nests are often open to the external air, but more generally deep in clefts and crevices or under stones and rocks. A covered site seems always to be preferred, but where the birds are numerous the less fortunate individuals have to be content with more open and exposed situations. In crowded colonies many nests are built close together, in some cases absolutely touching each other. The nest is a cup-shaped structure loosely put together and made of dry grass, twigs, stalks of plants, straws, bits of moss, and lined with finer grass, dry leaves, and feathers, although the latter material is sometimes omitted. Like the Starling the present species frequently drops an egg on the ground whilst feeding.

RANGE OF EGG COLOURATION AND MEASUREMENT: The eggs of the Rose-coloured Starling are from five to seven in number, and uniform pale bluish-white in colour, smooth and possessing considerable gloss. Average measurement, 1·1 inch in length, by ·82 inch in breadth. Incubation, performed by the female, who is said to sit very close, is presumed to last fourteen days.

DIAGNOSTIC CHARACTERS: Usually the eggs of the Rose-coloured Starling may be distinguished from those of the Common Starling by their much paler hue and slightly smaller size, but they require the most careful identification, and should always be well authenticated.

Family FRINGILLIDÆ. Genus LOXIA.
Sub-family *FRINGILLINÆ.*

AMERICAN WHITE-WINGED CROSSBILL.

LOXIA LEUCOPTERA, *Gmelin.*

(British : Rare nomadic autumn migrant.)

Number of broods unknown. Laying season, March and April.

BREEDING AREA: Northern Nearctic region. The American White-winged Crossbill probably breeds throughout Canada and Alaska in all suitable districts up to the limit of forest growth.

BREEDING HABITS: Even in the rigorous climate of British North America this White-winged Crossbill is a very early breeder, and begins nesting long before the snow is off the ground. But little has been recorded of the habits of this species. Its favourite breeding haunts are belts and plantations of fir and other coniferous trees. Whether it breeds in scattered colonies seems not to be known; whether it pairs annually or remains united for life is still undetermined. The nest is usually made in a pine or fir tree, twenty or more feet from the ground, and is described as a deep saucer-shaped structure made externally of spruce twigs and fibrous lichens, lined with hair and shreds of inner bark. A nest obtained by Dr. Adams at Fredericton in New Brunswick is said by Messrs. Baird, Brewer, and Ridgway to measure a little under four inches in external diameter, and the cavity to be two and a half inches across and one and a half inches in depth. The behaviour of the parents at the nest when disturbed is still undescribed.

RANGE OF EGG COLOURATION AND MEASUREMENT: The eggs of the American White-winged Crossbill are not known to exceed four in number. They vary from pale blue to white with a tinge of green or blue in ground colour with dark-brown and paler brown surface spots and specks, and underlying markings of a similar character, but violet-gray in colour. As a rule the markings are most thickly distributed on the large end of the egg, round which they generally form an irregular zone. Average measurement, ·78 inch in length by ·55 inch in breadth. Incubation presumably lasts fourteen days, but whether performed solely by the female is not known.

DIAGNOSTIC CHARACTERS: It is impossible to give any character by which the eggs of this species can be distinguished from those of the Greenfinch and the Common Crossbill. Unless they are thoroughly well identified and authenticated they are worthless as scientific specimens. The *locality* is of some small service in the matter of identification.

Family FRINGILLIDÆ.
Sub-family *FRINGILLINÆ.*

Genus LOXIA.

EUROPEAN WHITE-WINGED CROSSBILL.

LOXIA BIFASCIATA (*Brehm*).

(British : Rare nomadic autumn and winter migrant.)

Number of broods unknown. Laying season, apparently unknown.

BREEDING AREA: Northern Palæarctic region. The European White-winged Crossbill breeds in the pine forests of Northern Russia and Siberia, as far east as the Pacific coast, where it was observed by Middendorff in lat. 55°. Its southern breeding range is not very minutely determined, but certainly descends in the valley of the Yenesay as low as lat. 63°; and according to Dybowsky, includes the mountain forests near Lake Baikal.

BREEDING HABITS: It is not a little remarkable that the breeding habits of such a fairly well-known bird as the European White-winged Crossbill are almost completely unknown and undescribed. Its haunts are accessible, and have been well explored by many competent naturalists. The reason for this want of knowledge is probably due to the fact that the bird is a very early breeder, and has finished nesting before the northern forests are free from snow, or visited by travellers. A specimen of the nest of this Crossbill, which was sent to Mr. Dresser from Archangel, is described by that naturalist as smaller than that of the Common Crossbill and somewhat slighter in structure, which amounts to no description at all, and deserves to be pilloried as a fine example of slipshod work and opportunity neglected.

RANGE OF EGG COLOURATION AND MEASUREMENT: Precisely similar remarks apply to the description of the eggs of this Crossbill; for to state that they are smaller and rather darker in colour than those of the Common Crossbill is a most unsatisfactory statement. Period of incubation unknown.

DIAGNOSTIC CHARACTERS: In the absence of any trustworthy specimens of the eggs of this species, I am not prepared to say whether they may or may not be distinguished from those of allied species.

THE PARROT CROSSBILL (*Loxia pityopsittacus*), I may here remark, is merely a race of the Common Crossbill, not recognized, even sub-specifically, by recent writers.

Family FRINGILLIDÆ.   Genus LOXIA.
Sub-family FRINGILLINÆ.

PINE-GROSBEAK.

LOXIA ENUCLEATOR, *Linnæus.*

(British : Very rare nomadic winter migrant.)

Single Brooded. Laying season, late in May and in June.

BREEDING AREA : Northern Nearctic and Palæarctic regions. The Pine-Grosbeak is a circumpolar species, and breeds principally in the coniferous forests bordering the Arctic circle. In the Old World its breeding range extends from Norwegian Lapland across North Russia and Siberia to Kamtschatka, and south to the mountain pine forests near Lake Baikal. In the New World it extends from Alaska to Labrador, and probably includes the whole of Canada, even the southern provinces of Ontario and New Brunswick, in both of which the nest of this bird has been taken.

BREEDING HABITS: The favourite breeding haunts of the Pine-Grosbeak are the open spaces in the pine woods, and rough uneven country where the spruce firs and the larches occur in scattered groves or in single trees, and by the sides of the Arctic streams on the outskirts of the forests. Nothing appears to be known of the pairing habits of this species. Apparently it is not gregarious during the breeding season, but the available evidence points to a certain amount of sociability. John Wolley was the first naturalist to obtain any reliable information concerning the nest and eggs of the Pine-Grosbeak. He found it breeding near Muonioniska in Lapland in 1855 ; and in 1862 Wheelwright, better known as the "Old Bushman," procured a series of five nests near Quickiock in Northern Sweden, during the same year that he was fortunate enough to obtain the eggs of the Waxwing. The nest of the Pine-Grosbeak is usually made in a spruce fir, or less frequently in a birch, some ten or twelve feet from the ground, on a horizontal branch near the trunk. It is somewhat loosely put together, and made much on the same model as that of the Bullfinch or the Crossbill. Externally it is made of fir or birch twigs carelessly interlaced, and internally of roots, grass, and fine hair-like lichens. It is rather flat, and the finer materials project for some way over the foundation of twigs. During the nesting season this bird is shy and seclusive, especially the female, who when flushed from her charge makes little demonstration unless the eggs are much incubated.

RANGE OF EGG COLOURATION AND MEASUREMENT: The eggs of the Pine-Grosbeak are normally four in number, but occasionally three. They are clear greenish-blue in ground colour, handsomely spotted or speckled with rich blackish-brown and paler brown, and with underlying markings of lavender-gray. On some eggs the spots form an irregular zone round the larger end, many being frequently confluent, whilst on most the markings are chiefly about that part of the egg. An occasional streak of dark-brown sometimes occurs. On some eggs the markings are smaller—little more than specks—and more evenly dispersed over the entire surface. Average measurement, 1·0 inch in length, by ·72 inch in breadth. Period of incubation apparently unknown.

DIAGNOSTIC CHARACTERS: The eggs of the Pine-Grosbeak cannot readily be confused with those of any other British species, their large size readily identifying them amongst those of all other Finches reputed British.

Family FRINGILLIDÆ.   Genus CARPODACUS.
Sub-family FRINGILLINÆ.

SCARLET ROSE FINCH.

CARPODACUS ERYTHRINUS (*Pallas*).

(British : Very rare abnormal autumn migrant.)

Single Brooded. Laying season, May, June, and early July.

BREEDING AREA: Palæarctic region. The Scarlet Rose Finch breeds from the Baltic Sea across Europe and Asia to the Pacific. Its most westerly breeding-grounds are situated in Finland, the Baltic Provinces of Russia, and North-east Prussia (Silesia, Königsberg, Memel). Thence they extend eastwards into the Urals, the Caucasus, Asia Minor, Turkestan, Central Asia, Cashmere, and Mongolia to Kamtschatka, and northwards to about the latitude of the Arctic Circle.

BREEDING HABITS: The Scarlet Rose Finch arrives at its breeding grounds in Western Europe about the middle of May. Its winter quarters are in India! Whether this species pairs annually or remains united for life appears not to be known. It probably pairs every year. The favourite haunts of this bird are marshy forests and swamps with plenty of underwood. In Prussia its chief summer haunts are alder swamps; it also shows some partiality for dense willow thickets. During the earlier periods of the breeding season the male makes himself very conspicuous, sitting on the topmost twigs of the bushes and small trees, and uttering from time to time his singular song, which is described as a loud clear whistle, which arrests the attention at once. The Scarlet Rose Finch, if not gregarious, must at least be very social during the breeding season, for Herr E. Hartert records that in North-eastern Prussia seventeen nests were found in one wood during a single season. The nest is generally placed in a convenient crotch of some bush or low tree, and is of a flimsy character—a most unusual type of nest in this family of birds. It is somewhat deep and made of dry stalks of grasses, lined with horse-hair, beautifully rounded and neatly put together. Of the behaviour of the bird when disturbed at the nest I can find nothing recorded; but the female apparently is shy and retiring at this period.

RANGE OF EGG COLOURATION AND MEASUREMENT: The eggs of the Scarlet Rose Finch are from four to six in number; five is the average clutch. They are deep bluish-green in ground colour, spotted and speckled somewhat sparingly with rich blackish-brown and paler brown, and with underlying markings of gray. On many eggs the spots form an irregular zone round the larger end. Average measurement, ·8 inch in length, by ·59 inch in breadth. The period of incubation (which is performed by the female) appears to be unknown.

DIAGNOSTIC CHARACTERS: I know of no reliable character by which the eggs of the Scarlet Rose Finch can be distinguished from those of the Bullfinch. The spots may be darker and certainly fewer than on the eggs of the latter bird, but this test is worthless. The *nests* of the two birds cannot possibly be confused, however, even if the parents are not seen.

Family FRINGILLIDÆ.
Sub-family *FRINGILLINÆ.*
Genus SERINUS.

CANARY.

SERINUS HORTULANUS CANARIUS (*Linnæus*).

(British : Rare abnormal spring and autumn migrant.)

Double Brooded. Laying season, January to July, according to altitude.

BREEDING AREA : Portions of Cisatlantean sub-region of Palæarctic region. The Canary breeds and is a resident in the Canary Islands, Madeira, and the Azores. It is merely an island race of the Serin Finch.

BREEDING HABITS : Unfortunately we are in possession of little detailed knowledge concerning the reproduction of the canary in a wild state. The bird probably pairs annually, and there appears to be some evidence that it is to some extent social during the nesting season, and congregates into flocks of varying size after that period is passed and the young are reared. The breeding haunts of the Canary embrace a great variety of scenery. The bird not only breeds in the gardens and country near the towns, but is found in the mountain pine forests up to an altitude of six thousand feet. Like so many of our own Finches at home, the Canary usually selects a nesting site in an evergreen tree or bush, sometimes at a considerable height from the ground, and seldom less than six or eight feet from it. The nest is made externally of the stems of plants and coarse grass, and lined with finer grass and vegetable down. I find little or nothing recorded of the behaviour of the parent birds at the nest when disturbed.

RANGE OF EGG COLOURATION AND MEASUREMENT : The eggs of the Canary are from four to six in number.

They are pale bluish-green in ground colour, spotted and speckled with reddish-brown, and with underlying markings of gray. On many eggs the markings are distributed in an irregular zone round the large end. Clutches are sometimes found pure white. Average measurement, ·8 inch in length, by ·55 inch in breadth. Incubation, performed by the female, lasts twelve or thirteen days.

DIAGNOSTIC CHARACTERS: I know of no reliable character which will serve to distinguish eggs of the Canary from those of the Linnet and the Greenfinch. The style of the nest is of some service, but the eggs of this species require very careful identification and strict authentication to be of any scientific value.

Family FRINGILLIDÆ.   Genus SERINUS.
Sub-family FRINGILLINÆ.

SERIN FINCH.

SERINUS HORTULANUS, *Koch.*

(British : Rare abnormal spring and autumn migrant.)

Single Brooded. Laying season, March, April, and May.

BREEDING AREA: South-west Palæarctic region. The Serin Finch is confined during the breeding season to South-western Europe, Asia Minor, and North-western Africa. It breeds in South Prussia, Luxemberg, Central and Southern France, the Spanish peninsula, Switzerland, Italy, Austria, Turkey, Greece, and Asia Minor, and south of the Mediterranean in Algeria and Morocco.

BREEDING HABITS: The haunts of the Serin during the breeding season are orchards, gardens, pleasure-grounds and vineyards, as well as the mountain thickets and conifer woods. It is by no means a retiring species, and I have repeatedly seen it within arm's length of houses. Of its pairing habits nothing appears to have been recorded. It is not gregarious during the breeding season, but is certainly socially inclined, and several pairs frequently nest in one orchard or garden. During the early part of the breeding season, the male makes himself very conspicuous, perching on the topmost branch of a tree, or on the extremity of a large branch at some distance from the ground, and uttering his clear sweet little song. The female is much more seclusive, and during the breeding season keeps chiefly to the cover. The nest of the Serin is generally built in a dense bush or at a moderate height in a fruit or other tree. I have noticed the partiality of this bird for oleander bushes and pomegranates. The nest is small and neat, made externally of roots and dry stalks cemented with bits of lichen, cocoons, or spiders' webs, and lined with vegetable down, and sometimes wool and moss. Occasionally hair and feathers are used in the lining.

RANGE OF EGG COLOURATION AND MEASUREMENT: The eggs of the Serin are four or five in number. They vary from pale bluish-green to greenish-white in ground colour, spotted and blotched with light-brown and dark reddish-brown, and with underlying markings of paler brown and gray. Occasionally a few dark-brown streaks occur, chiefly at the large end of the egg, round which most of the markings are distributed, frequently in the form of a zone. On some eggs most of the markings partake more of the character of blotches than spots. Average measurement, ·62 inch in length by ·47 inch in breadth. Incubation, performed chiefly if not entirely by the female, lasts, according to Bechstein, thirteen or fourteen days.

DIAGNOSTIC CHARACTERS: It is impossible to give any character by which the eggs of the Serin can be distinguished from those of the Goldfinch and the Siskin. They require the most careful identification and reliable authentication to render them of any scientific value.

Family FRINGILLIDÆ.
Sub-family *FRINGILLINÆ*.

Genus FRINGILLA.

BRAMBLING.

FRINGILLA MONTIFRINGILLA, *Linnæus*.

(British: Common autumn migrant.)

Double Brooded. Laying season, May and June.

BREEDING AREA: Northern Palæarctic region. The Brambling breeds in the pine and birch forests of the Arctic regions from the Atlantic to the Pacific. In Scandinavia it does not appear to breed south of lat. 60°, but in the far East in the valley of the Amoor, where the mean summer temperature is lower, it does so as far south as lat. 50°.

BREEDING HABITS: Although the Brambling is obviously very closely allied to the Chaffinch, the habits of the two species, especially during the breeding season, differ considerably. The Brambling is much more of an Arctic bird, and even in the southern limits of its breeding area its ally the Chaffinch becomes decidedly rare. During the nesting season the Brambling is, if not exactly so gregarious as during winter, at least social, and numbers of pairs generally breed in company. The favourite breeding grounds of the Brambling are birch forests, and open straggling woods composed of firs and birches, with a few larches and alders intermingled. Birch, willow, and alder thickets are also frequented. It is probable that the Brambling pairs annually, and from what I have observed I should think that this event takes place in spring before the birds leave their winter quarters. Amongst large timber the nest is generally made at heights varying from ten to thirty feet above the ground, but amongst smaller trees and in thickets it is placed at a much lower elevation. A favourite site is in a crotch where some branch joins the trunk, or in a fork of a limb at some distance from the stem. In bushes several small branches often support the nest. This is cup-shaped, but nothing near so neat a structure as that of the Chaffinch, and is more loosely put together and larger. Externally it is composed of moss, lichens, and birch bark studded with cobwebs and vegetable down, which serve to bind the other materials together, and lined with fine dry grass and quantities of feathers. Whether the female is the sole builder, as is the case with the Chaffinch, appears not to have been determined, and the actions of the birds when disturbed at their nesting-places appear also to be undescribed.

RANGE OF EGG COLOURATION AND MEASUREMENT: The eggs of the Brambling are from five to seven in number, six being an average clutch. They are bluish-green in ground colour, spotted and speckled with under-lying markings of pale reddish-brown and surface spots of very dark brown, and generally more or less clouded and suffused with irregular blurred blotches of very pale brown. The dark spots are often very rotund, and surrounded by paler discs of colour. One type is almost devoid of these dark surface spots; another has a circular patch of gradated colour like a cap over the larger end of the egg; another type is almost uniform pale-green without markings. Average measurement, ·8 inch in length by ·6 inch in breadth. Incubation is performed by the female, but the length of the period appears to be unknown; probably it is similar to that of the Chaffinch, and lasts from twelve to fourteen days.

DIAGNOSTIC CHARACTERS : The only eggs with which those of the Brambling can be confused are those of the Chaffinch, and unfortunately no reliable character can be given to distinguish them. As a rule the eggs of the Brambling are more vivid green in ground colour, more clouded, and the spots are not so large and distinct.

Family FRINGILLIDÆ.   Genus LINOTA.
Sub-family *FRINGILLINÆ*.

MEALY REDPOLE.

LINOTA LINARIA (*Linnæus*).

(British: Autumn migrant.)

Single Brooded. Laying season, May and June.

BREEDING AREA : Northern Nearctic and Palæarctic regions. The Mealy Redpole breeds principally in the birch regions, at or near the limits of forest growth in Europe, Asia, and North America. Its southern range is not very accurately determined, inasmuch that the bird has been confused with the Lesser Redpole, especially in the Alpine districts of Europe.

BREEDING HABITS : So far as can be gathered from the somewhat meagre details on record regarding the nidification of the Mealy Redpole, there is little if any difference between this species and the, to British ornithologists, better known Lesser Redpole. The Mealy Redpole probably pairs annually. The favourite breeding haunts of this species appear to be birch woods and thickets, and in the high north near the limits of forest growth, where these trees dwindle down into mere bushes, they still seem to be preferred, but stunted willow thickets are often selected. The nest is rarely if ever built at any great distance from the ground, even where the timber is of fair size, and in districts where the growth is stunted it is seldom more than four feet above it. Sometimes it is placed almost on the ground in a tuft of herbage. The nest is made externally of twigs, roots, moss, and dry grass, lined with vegetable down of some kind, principally from willow catkins and cotton grass, or feathers, and hair. The lining varies a good deal according to locality and availability. It is not known that the conduct of the female at the nest differs from that of the Lesser Redpole.

RANGE OF EGG COLOURATION AND MEASUREMENT: The eggs of the Mealy Redpole are five or six in number. They are blue in ground colour, but on some a greenish tinge predominates, on others it is scarcely perceptible. They are spotted and speckled with purplish-brown, and with underlying markings of paler brown and gray. Most of the markings are on the larger end of the egg; and on some eggs a few streaks of dark brown occur. A rare variety is very sparingly marked. Average measurement, ·7 inch in length, by ·5 inch in breadth. Incubation is probably performed by the female alone, but the duration of the period is unknown. It is probably the same as that of the Lesser Redpole, namely, fourteen days.

DIAGNOSTIC CHARACTERS: The eggs of the Mealy Redpole are perceptibly larger than those of the Lesser Redpole, but do not differ in colour or character of markings. The locality of the eggs is a fairly reliable means of identification.

Family FRINGILLIDÆ.   Genus LINOTA.
Sub-family FRINGILLINÆ.

GREENLAND REDPOLE.

LINOTA LINARIA HORNEMANNI, *Holböll.*
(British : Rare abnormal winter migrant.)
Single Brooded. Laying season, June.

BREEDING AREA : North-east of Nearctic region, and extreme north-west of Palæarctic region. The Greenland Redpole breeds in Spitzbergen, Iceland, Greenland (north of lat. 69°), and possibly in Canada.

BREEDING HABITS : But little has been recorded of the nidification of this race of the Mealy Redpole, but probably it differs little, if at all, from that of the preceding bird, due allowance being made for local conditions. The nest so closely resembles that of the Mealy Redpole that a detailed description is unnecessary, and it appears to be made in precisely similar situations—in stunted bushes, or even on the ground above the limits of forest growth, in small trees in lower latitudes.

RANGE OF EGG COLOURATION AND MEASUREMENT : The eggs of the Greenland Redpole are five or six in number. They range from bluish-green to greenish-blue in ground colour, spotted and speckled with dark purplish-brown, and with underlying markings of paler brown and gray : a few streaks occasionally occur. The markings are often distributed in a zone round the larger end of the egg, and are almost invariably most numerous on that portion. Average measurement, ·75 inch in length by ·54 inch in breadth. Incubation, probably performed by the female ; the duration of the period is unknown, but is probably the same as that of the allied races.

DIAGNOSTIC CHARACTERS: The eggs of the Greenland Redpole are the largest of the three races of Redpole that occur within the area of the British Islands, but do not differ in colour or character of markings. The locality is a tolerably safe guide to their identification, if it can be relied upon.

Family FRINGILLIDÆ. Genus CALCARIUS.
Sub-family EMBERIZINÆ.

LAPLAND BUNTING.

CALCARIUS LAPPONICUS (*Linnæus*).

(British : Rare abnormal autumn migrant.)

Single Brooded.   Laying season, June.

BREEDING AREA : Northern Nearctic and Palæarctic regions. The Lapland Bunting breeds on the tundras and barren grounds above the limit of forest growth in Europe, Asia, and North America, and in a similar region at high elevations in various parts of the Dovre Fjeld in Norway. Singularly enough it is not known to breed in Iceland, Spitzbergen, or Nova Zembla, but is common across Northern Europe and Asia, in Europe extending as far north as continental land exists, but in Asia not apparently beyond about lat. 73°. In Greenland it breeds as far north as lat. 70°, and is said to be abundant in the Arctic regions of the New World from Alaska eastwards.

BREEDING HABITS : The Lapland Bunting is another typical boreal species, only breeding on the treeless wastes of the north, above the limits of forest growth— on the Arctic prairies or flower-decked moors that stretch away beyond the forests to the shores of the Polar seas.

Here, however, the bird appears to show partiality for certain haunts. Its favourite nesting places are the swampy grounds and marshes studded with hassocks of grass and flowers which afford dry accommodation for its home, and relieved here and there with stunted willows and birches, which afford the bird a resting-place. The Lapland Bunting pairs annually, and the males arrive some little time before the females at the breeding grounds. Upon the arrival of the latter but little time is wasted before the duties of the year commence. The birds are not gregarious during the breeding season, but numbers of pairs may be found breeding within a small area. The cock-birds are very musical during the early days of the breeding period, warbling whilst soaring in the air after the manner of a Pipit. The nest is generally placed in a hollow in the side of one of the grassy hummocks or tufts of herbage, and is composed externally of dry grass, scraps of moss and roots, and lined with finer grass and an abundance of feathers. The female is a rather close sitter, often betraying the nest by flying up at the observer's feet, and when disturbed evinces considerable anxiety, especially if the eggs be much incubated.

RANGE OF EGG COLOURATION AND MEASUREMENT: The eggs of the Lapland Bunting are from four to six in number. They vary in ground colour from pale olive or grayish-brown to chocolate or sienna-brown, blotched, spotted, and streaked with brown of various shades. Although subject to considerable variety, the eggs in each clutch are pretty uniform amongst themselves. Most of the larger markings are pale and underlying ones, the surface spots being much less numerous, darker and richer, and often taking the form of streaks and hair-like lines intermingled with blotches. Average measurement, ·82 inch in length, by ·6 inch in breadth. Incubation is chiefly performed by the female, but the duration of the period is unknown.

DIAGNOSTIC CHARACTERS: The eggs of the Lapland Bunting may be confused with those of the Tree Pipit and the Red-throated Pipit, and as the measurements of the eggs of all three species intergrade no reliable point of distinction can be given. It may be remarked, however, that the Tree Pipit does not intrude upon the breeding area of the Lapland Bunting, so that locality is sufficient to identify the eggs of the latter; whilst in both these Pipits the nest is never lined with feathers, but the Bunting's invariably is.

Family FRINGILLIDÆ.   Genus EMBERIZA.
Sub-family EMBERIZINÆ.

BLACK-HEADED BUNTING.

EMBERIZA MELANOCEPHALA, *Scopoli.*

(British : Very rare abnormal spring and autumn migrant.)

Single Brooded.   Laying season, latter half of May and early in June.

BREEDING AREA: South-western Palæarctic region. The Black-headed Bunting breeds in Northern Italy, Dalmatia, Turkey, Greece, the Danubian Provinces, Southern Russia, the Caucasus, Asia Minor, Palestine, and Northern Persia.

BREEDING HABITS: The Black-headed Bunting does not arrive at its western breeding grounds until the end of April. It is in no way gregarious during the summer months, not even social, but in districts where it is common many pairs may be found nesting within a very small area. This bird pairs annually, and the evidence seems to suggest that the event takes place before departure from the winter quarters in India. The favourite haunts of this Bunting during the breeding season appear to be the lower mountain slopes and uneven plains at no great distance from the sea. It does not, however, go so far up the mountains as the pine regions. In these districts it frequents gardens, vineyards, olive-groves, and the open rough country clothed with a luxuriant growth of clematis, fig, almond, pomegranate, oleander and briars, and studded with rock fragments. Here the Black-headed Bunting is just as pertinacious as its cousin the Yellow Bunting is in our English fields, and the males may be seen sitting conspicuously on the tall bushes and trees singing in rivalry. The nest of this species, built at no great distance from the ground, and even on the ground itself, is made in a dense bush or amongst the wild untrimmed growth of briar, clematis, or vine, whilst in gardens rows of peas or beans are generally selected. The nest, which is cup-shaped, is made externally of coarse grass, the dry stems of plants, with the seeds or dry flowers attached, and dead leaves, and internally of dry grass bents, roots, and hair, although the latter material is not always used, probably because it is not available. The nest is a remarkably easy one to find, and very often the obtruding loquacious male betrays its whereabouts. The female, in her much more sombre dress, is quiet and unassuming, and sits closely.

RANGE OF EGG COLOURATION AND MEASUREMENT: The eggs of the Black-headed Bunting are without the characteristics of most Buntings' eggs. They are from four to six in number, four or five being most frequently found. The ground colour is very pale greenish-blue;

the surface spots are brown, the underlying ones are gray. The eggs vary considerably in size and character of markings. On some eggs the latter are few, large, and irregular, on others small and evenly sprinkled over the entire surface, whilst on others the small spots predominate, intermingled here and there with larger markings. As a rule most of the spots are distributed over the larger end of the egg, and sometimes, being confluent, form an irregular zone. Average measurement, ·85 inch in length, by ·7 inch in breadth. Incubation is performed by the female, but the duration of the period is apparently unknown.

DIAGNOSTIC CHARACTERS: Some of the eggs of this species closely resemble those of the White and Pied Wagtails, but the locality and structure of the nest is a pretty safe guide. They cannot be confused with those of any other Bunting, except perhaps with those of *Emberiza luteola*, a species however which does not breed in the same area.

Family FRINGILLIDÆ.   Genus EMBERIZA.
Sub-family *EMBERIZINÆ*.

ORTOLAN BUNTING.

EMBERIZA HORTULANA, *Linnæus*.

(British : Abnormal spring and autumn migrant.)

Single Brooded. Laying season, latter half of May and early June.

BREEDING AREA: Western Palæarctic region. The Ortolan Bunting breeds throughout continental Europe in all suitable localities, extending as far north as the

Arctic circle in Scandinavia, but not beyond lat. 57° in the Urals. It also breeds in Asia Minor, Palestine, Persia, Turkestan, and Siberia, as far east as the valley of the Irtish and the Great Altai. It is said also to breed sparingly in North-west Africa, but details of its distribution are wanting.

BREEDING HABITS: The Ortolan Bunting arrives at its more southerly breeding grounds in Europe during the first half of April, but it is about a month later in the northern districts. The haunts of this species are in cultivated localities as well as in mountain forest areas, and the more open country in wilder districts. Of its pairing habits we possess no information, but doubtless the bird mates annually. The nest, made on the ground, is generally concealed amongst herbage or under the shelter of a low bush, frequently amongst the grass in an open field or on a hassock of drier vegetation in a wet meadow or clearing of the forest. It is cup-shaped and somewhat loosely made, composed externally of dry grass, stalks of plants and coarse roots, and lined with fine roots and hair. The female, like most ground-building birds, is a close sitter, and when flushed makes little or no demonstration at the nest. This species is neither gregarious nor social during the breeding season, living in scattered pairs.

RANGE OF EGG COLOURATION AND MEASUREMENT: The eggs of the Ortolan Bunting are usually five, but four and six are sometimes found. They vary in ground colour from bluish-white to very pale purplish-gray, spotted and blotched, and more rarely streaked with very dark purplish-brown and paler brown, and with underlying markings of gray. As a rule the spots are large and round, but the streaks and lines are short and not continuous. As usual most of the colouring is on the larger end of the egg, and frequently forms a zone.

Average measurement, ·81 inch in length by ·62 inch in breadth. Incubation is performed chiefly by the female, but the duration of the period appears not to have been observed.

DIAGNOSTIC CHARACTERS: The eggs of the Ortolan Bunting most closely resemble those of the Reed Bunting, but are larger, not so clouded in appearance, and the markings partake more of the character of spots than interlaced streaks.

Family FRINGILLIDÆ.   Genus EMBERIZA.
Sub-family *EMBERIZINÆ*.

LITTLE BUNTING.

EMBERIZA PUSILLA, *Pallas.*

(British: Very rare abnormal autumn migrant.)

Single Brooded. Laying season, June and beginning of July.

BREEDING AREA: Northern Palæarctic region. The Little Bunting breeds from the eastern shores of the White Sea across Northern Europe and Asia to the Tchuski Land and the valleys of the Lower Amoor. It is a thoroughly boreal species, and seldom breeds south of the Arctic circle, except at high elevations, as for instance on the mountains of Eastern Siberia and in the Baikal area. It was met with in the valley of the Petchora as far north as lat. 68°; in the valley of the Yenesay up to lat. 71°.

BREEDING HABITS: The Little Bunting, like most other Arctic birds, is a late migrant, not reaching its breeding grounds in the Arctic regions before the end of May or the first week in June, according to the state of the season. If it does not actually migrate in pairs it soon mates after its arrival in its summer haunts. The males are persistent singers, as is usual in this group of birds. The favourite breeding haunts of the Little Bunting are the small pine and fir woods intermixed with birches and alders, and where the undergrowth is fairly dense. Although Schrenck took the nest of this Bunting on the Lower Amoor, and Middendorff obtained others on the banks of the Boganida, these naturalists do not give many details of their discovery, and by far the best account of the nidification of this species is that written by Mr. Seebohm, who has added so much to our information of the habits of birds in the Arctic regions, and whose discoveries, made both by himself and when accompanied by Mr. Harvie-Brown, are tolerably well known to most ornithologists. He found the Little Bunting very common in Siberia, in the valley of the Yenesay, and discovered the first nest on the 23rd of June; and between that date and the 6th of July succeeded in taking three others. They were made on the ground amongst the grass and moss and fallen leaves, and were hollows amongst the dead leaves, moss and grass lined with dry grass, and in two instances reindeer hair. The nest found by Schrenck was also on the mossy ground, and made of dry grass and the needle-like leaves of the fir. The sitting bird is described as being remarkably tame when flushed from the eggs.

RANGE OF EGG COLOURATION AND MEASUREMENT: The eggs of the Little Bunting are from four to six in number. Those that I have examined vary in ground colour from pale grayish-olive to pale reddish-brown, spotted and blotched with dark olive-brown, and with underlying markings of paler brown. The larger markings are irregular in shape, but many of the smaller ones are round. Average measurement, ·73 inch in length by ·57 inch in breadth. Incubation is performed chiefly by the female, but the duration of the period is apparently unknown.

DIAGNOSTIC CHARACTERS : The eggs of the Little Bunting cannot readily be confused with those of any other species ; they most nearly resemble those of the Corn Bunting, but are of course much smaller.

Family ALAUDIDÆ.   Genus OTOCORIS.

SHORE-LARK.

OTOCORIS ALPESTRIS (*Linnæus*).

(British : Nomadic autumn migrant.)

Partially Double Brooded. Laying season, middle of May to end of June.

BREEDING AREA : Northern Nearctic and Northern Palæarctic regions. The Shore-Lark is another strictly Arctic species, breeding on the tundras and barren lands above the limits of forest growth in both hemispheres. In the Old World it breeds in the north-east of Norway, and on the fells as far south as lat. 67°; thence it extends eastwards across Russian Lapland, the province of Archangel, including Nova Zembla, and the north of Siberia to Bering Strait. In the New World it breeds above forest growth from Alaska to Greenland ; it is said that a few pairs remain to breed near Toronto (*fide* Raine).

BREEDING HABITS: The Shore-Lark, like most nomadic migrants, appears early at its breeding grounds, reaching them the moment sufficient snow has melted to allow of its obtaining food. Owing to its partiality for sandy ground, bare rocky hills, and the sloping banks of clay that confine the Arctic rivers in many places, the Shore-Lark is certainly a local bird. It is especially fond of such portions of the tundras and barren grounds as are sandy and clothed with scrub, eschewing the more marshy portions of these Arctic moors. It migrates north in spring in flocks of varying size, but these disband after pairing, and during the breeding season it lives in scattered pairs. The male is a most persistent singer, warbling whilst soaring after the manner of a Sky-Lark, or when sitting on some building, but this species rarely if ever perches in trees. The nest of the Shore-Lark is invariably made upon the ground, and, as is the case with other Larks, usually in a hollow of some kind. Very bare and exposed situations are sometimes selected, as for instance amongst loose stones, or even on a frequented path, but more usually the shelter of a tuft of herbage is sought. The nest, which is cup-shaped, is made externally of dry grass, bits of moss, and the stalks of plants, and lined with vegetable down of some kind, or reindeer hair if it can be obtained. It is said that feathers are also sometimes used. It is somewhat carelessly put together like the nests of most of the Larks. The female is a close sitter, remaining until almost trodden upon ere she rises.

RANGE OF EGG COLOURATION AND MEASUREMENT: The eggs of the Shore-Lark are from three to five in number, four being the average clutch. They vary in ground colour from greenish-white to brownish-white, indistinctly mottled, freckled, and spotted with olive-brown, and with underlying markings of paler brown and gray. Occasionally a few hair-like very dark brown streaks and spots at the large end, more rarely over the entire surface, occur. As a rule the surface markings are so numerous that most of the ground colour is concealed, and the underlying markings are inconspicuous. On some eggs the markings are unusually dense round the larger end, forming a cap or a zone. A rarer type displays more ground colour, and the underlying markings are very conspicuous and deep violet-gray. Average measurements, ·9 inch in length by ·63 inch in breadth. Incubation is performed by the female, but the duration of the period is apparently unknown.

DIAGNOSTIC CHARACTERS: It is impossible to give any character by which the eggs of the Shore-Lark may be distinguished from those of the Sky-Lark. On an average they are slightly smaller, and perceptibly more olive in general appearance. The Sky-Lark, however, does not frequent the breeding area of the Shore-Lark, and never lines its nest so warmly.

Family ALAUDIDÆ. Genus MELANOCORYPHA.

CALANDRA LARK.

MELANOCORYPHA CALANDRA (*Linnæus*).

(British: Very rare abnormal autumn migrant.)

Single Brooded. Laying season, April to June, according to locality.

BREEDING AREA: South and South-west Palæarctic region. The Calandra Lark breeds in all suitable localities throughout the basin of the Mediterranean, but not north of lat. 46° in the west although slightly beyond in the east. It breeds in Africa north of the Atlas, and thence through Asia Minor into Persia and Turkestan to the western slopes of the Great Altai.

BREEDING HABITS: In many of its habits the Calandra Lark closely resembles the Sky-Lark, but it is more fastidious in its choice of a haunt. It loves warm sandy soils, and is especially fond of steppe country in districts where cereals are largely grown. I remarked its preference for such country especially in Algeria. During the breeding season the males are for ever fluttering into the air and dropping down again into the cover, sometimes singing, sometimes merely uttering their liquid call-note. This Lark is gregarious enough in winter, but during the nesting season lives in scattered pairs, which in some localities are often very thick on the ground. Of its pairing habits nothing definite appears to be known. I am inclined to think that it pairs annually, and at this season the males may be seen chasing each other and toying with the females. The nest is always made upon the ground in a slight hollow of some kind, and well concealed amongst growing corn or other herbage. It is a loosely made structure, composed externally of coarse dry grass, roots, and stalks of plants, and lined with finer grass and roots. It is a very close sitter, and when flushed from the nest flies right away at once with no demonstration of anxiety.

RANGE OF EGG COLOURATION AND MEASUREMENT: The eggs of the Calandra Lark are four or five in number. They vary in ground colour from grayish-white to white with a very perceptible yellowish tinge, thickly mottled and freckled with olive-brown, and with underlying markings similar in character of violet-gray. The markings are fairly well defined and close together, but a considerable amount of ground colour is visible. As a rule the spots are most numerous on the larger end of the egg, and sometimes form a cap or irregular zone.

Average measurement, ·95 inch in length by ·7 inch in breadth. Incubation is performed by the female, but the duration of the period is unknown.

DIAGNOSTIC CHARACTERS: It is impossible to give any character by which the eggs of the Calandra Lark can be distinguished from those of the White-winged Lark, but as a rule they are smaller. From those of the Sky-Lark they may be separated by their more pyriform shape and slightly larger size; the markings are also more scattered, and much more clearly defined.

Family ALAUDIDÆ. Genus MELANOCORYPHA.

WHITE-WINGED LARK.

MELANOCORYPHA SIBIRICA (*Gmelin*).

(British : Very rare abnormal autumn migrant.)

Single Brooded. Laying season, late in April and in May.

BREEDING AREA: South-central Palæarctic region. The White-winged Lark has a somewhat restricted range, and breeds in the extreme south-east of Russia from the province of Stavrapol, north of the Caucasus, across the steppes of Astrakhan and Saratov to Orenberg. Eastwards in Asia it breeds upon the Barabinska and Kirghiz steppes to as far east as the Great Altai.

BREEDING HABITS: The favourite haunts of the White-winged Lark are steppes bare of trees and open grassy plains. In its habits, so far as they are known, it very closely resembles the preceding species. In the pairing season it may be seen constantly soaring in short flights to sing. In the winter it is gregarious, but during the breeding season appears to live in scattered pairs. Very little is known of the nidification of this species. The nest is said to be placed upon the ground amongst herbage, either cultivated or wild, and to be made of grass, the finer kinds being reserved for the lining. Of its actions at the nest nothing has been recorded.

RANGE OF EGG COLOURATION AND MEASUREMENT: The eggs of the White-winged Lark are four or five in number, sometimes, it is said, only three. They are grayish or yellowish-white in ground colour, freckled and mottled with brown, and with underlying markings of lilac-gray. As a rule the markings are large and distinct, and generally distributed over the surface of the egg, but on some they are more confined to the larger end of the egg, where they frequently form a zone. The underlying markings are also very conspicuous and numerous. Average measurement, ·95 inch in length by ·65 inch in breadth. Incubation appears to be performed by the female, but the duration of the period is unknown.

DIAGNOSTIC CHARACTERS: Unfortunately the eggs of the White-winged Lark cannot with certainty be distinguished from those of the Calandra Lark; their more pyriform shape, larger size, and more scattered markings serve to distinguish them from those of the Sky-Lark.

Family ALAUDIDÆ.   Genus CALANDRELLA.

SHORT-TOED LARK.

CALANDRELLA BRACHYDACTYLA (*Leisler*).

(British : Rare abnormal autumn migrant.)

Single Brooded. Laying season, May and June.

BREEDING AREA : South-west Palæarctic region. The Short-toed Lark breeds in Central and Southern France, the Spanish Peninsula, Southern Germany, Italy, Turkey, Greece, the Danubian Provinces, and Southern Russia ; it also breeds on the various islands in the Mediterranean, in North-west Africa, and on the Canaries. Eastwards it breeds in Asia Minor, Palestine, Persia, Turkestan, and South-west Siberia as far east as the province of Semipolatinsk.

BREEDING AREA : The Short-toed Lark is another bird of the steppes, especially such as are of a sandy character. Its favourite breeding haunts are open plains and grassy downs, large fields of cereals and rough uncultivated lands often some distance above sea-level. It is a very obtrusive bird, incessantly taking songflights, and it has been heard even to warble on the ground. Although gregarious in winter it lives in scattered pairs during the breeding season. The nest is invariably made upon the ground, either amongst herbage or beneath the shade of a little bush, or even a clod of earth or dry dung, or a tuft of grass. It is usually made in a little hollow, either scraped out by the parent bird or selected ready for the purpose. This nest is cup-shaped, and made externally of dry grass, roots, and stalks of small plants, and lined with finer grass and roots. In some districts the nests appear to be much more elaborately made than in others, being lined with vegetable down and a few feathers. An instance is on record where a piece of white sail-cloth was inserted amongst the lining material. Curiously enough these warmer nests are recorded from the valley of the Danube and Northern Spain, and not from the cooler portions of the range of this species. The bird is a close sitter and undemonstrative.

RANGE OF EGG COLOURATION AND MEASUREMENT: The eggs of the Short-toed Lark are four or five in number, but it is said three are occasionally found. They are yellowish-white in ground colour, freckled with grayish-brown, and with underlying markings of lilac-gray. As a rule the profusion of the spotting almost obliterates the ground colour, but there are types in which the markings are somewhat scattered. Occasionally the markings are intensified, and form a zone round the larger end of the egg: they vary considerably in size. Average measurement, ·78 inch in length by ·57 inch in breadth. Incubation is performed by the female, but the duration of the period is unknown.

DIAGNOSTIC CHARACTERS: The comparatively small size and yellowish tinge serve to distinguish the eggs of the Short-toed Lark from other British species, but no character is known by which they may be separated from those of various closely allied species and races.

Family ALAUDIDÆ.   Genus GALERITA.

CRESTED LARK.

GALERITA CRISTATA (*Linnæus*).

(British : Rare abnormal autumn migrant.)

Single Brooded probably. Laying season, end of March to early June, according to latitude.

BREEDING AREA: South-west Palæarctic region. The typical form of the Crested Lark breeds in Central and Southern Europe from about lat. 60° down to the Mediterranean. It is rare in the extreme northern limits of its distribution, but south of the Baltic may be said to be common in all localities suited to its requirements.

BREEDING HABITS : The Crested Lark has a strong partiality for loose sandy soils, and its favourite breeding grounds are sandy heaths, and fields, and rough unenclosed lands, although in Algeria I met with it (*G. cristata magna*) in the Atlas at an elevation of five thousand feet on the rough stony hill-sides. It is gregarious to a certain extent during winter, but passes the summer in scattered pairs without any perceptible approach to sociability. The males are persistent singers during the early part of the breeding season, warbling in a Pipit-like way in the air as well as on the ground, or whilst perched on a bush or a telegraph wire. The nest of the Crested Lark is almost invariably made upon the ground, although instances are on record where it has been found on old earth-walls or amongst the thatch of sheds in the fields. The favourite sites are amongst the herbage of the fields, even on fallows, but in wilder districts the nest is frequently placed under a bush or amongst stones. It is composed externally of dry grass, stalks of plants, straws, and roots, and lined with finer grass, roots, and sometimes hair. The female is said to do the building, the male collecting the materials, and the nest is loosely made. The bird is a close sitter, and when disturbed flies away with no demonstration of alarm.

RANGE OF EGG COLOURATION AND MEASUREMENT: The eggs of the Crested Lark are four or five in number. They vary in ground colour from pale yellowish-white to white with a faint tinge of green or blue, mottled, freckled, and blotched with various shades of olive-brown, and with underlying markings of gray. Several distinct types are noticeable. One type has the markings dark and clearly defined, the gray spots very conspicuous, and both classes of spots evenly distributed. Another has the markings dark and very minute dusted over the surface, but most numerous at the larger end, where they frequently form a zone. A third is so closely mottled as to hide almost all trace of the pale ground colour. Average measurement, ·95 inch in length by ·68 inch in breadth. Incubation, performed by the female, lasts fourteen days.

DIAGNOSTIC CHARACTERS: The eggs of the Crested Lark, as I pointed out ten years ago in the *History of British Birds*, very closely resemble those of the Wood Lark, and I am unable to give any reliable character by which they may be separated. On an average those of the present species are larger, bulkier, and greener.

Family MOTACILLIDÆ. Genus ANTHUS.

ALPINE PIPIT.

ANTHUS SPIPOLETTA (*Linnæus*).

(British : Rare abnormal autumn migrant.)

Double Brooded. Laying season, from end of April to June.

BREEDING AREA : South-west Palæarctic region. The Alpine Pipit breeds locally on the mountains of Europe eastwards to Persia and Baluchistan. It is not known to breed in Scandinavia, but does so on the mountains of Central Europe, the Alps, the Pyrenees, the higher Spanish ranges, the Urals (as far north as lat. 64°), and the Caucasus. Eastwards it breeds on the highlands of Persia, Baluchistan, probably Afghanistan, and Turkestan as far east as the Great Altai.

BREEDING HABITS : There is much similarity between the habits of the Alpine Pipit and those of the Rock Pipit, only one bird loves a mountainous habitat and the other a littoral one. Both are rock-haunting species, only the Alpine Pipit, as its name implies, is a dweller on the mountains above the limits of forest growth during the breeding season. Its favourite breeding places are the swampy spots surrounded by rocks and clothed with a brilliant array of alpine blooms. Its habits do not differ in any important respect from those of allied and more generally distributed species. It pairs annually, but at what season we have no information. Although gregarious during winter, it lives in scattered pairs during the breeding season. The nest is placed either upon the ground, amongst loose stones, in a rock crevice, or beneath the shelter of an alpine bush or tuft of herbage, and is cup-shaped, composed externally of dry grass, moss, straws, and roots, and lined with fine grass, roots, hair, and, it is said, with wool or even feathers—which seems to show an adaptation to the low and probably varying temperature of the bird's haunts. When flushed from the nest this Pipit often betrays anxiety by flitting restlessly about and uttering its plaintive call-note.

RANGE OF EGG COLOURATION AND MEASUREMENT: The eggs of the Alpine Pipit are four or five in number, generally the latter. They vary in ground colour from bluish or greenish-white to creamy-white, mottled, freckled, and spotted with olive or purplish-brown, and with underlying markings of gray, and occasionally streaked with very dark brown. As is usual with the eggs of Pipits, we find several well-marked types, the most pronounced being olive, brown, and reddish-purple. As a rule the markings are small, and conceal most of the ground colour, whilst zones or caps often occur. Average measurement, ·85 inch in length by ·63 inch in breadth. Incubation, performed chiefly by the female, lasts thirteen or fourteen days.

DIAGNOSTIC CHARACTERS: It is impossible to give any character by which the eggs of the Alpine Pipit can be distinguished from those of the Rock Pipit, but the locality is quite sufficient to identify them.

Family MOTACILLID.E.   Genus ANTHUS.

TAWNY PIPIT.

ANTHUS CAMPESTRIS (*Linnæus*).

(British : Abnormal autumn migrant.)

Single Brooded.   Laying season, May and June.

BREEDING AREA : South-west Palæarctic region. The Tawny Pipit breeds throughout Europe in suitable localities south of about lat. 57°. It breeds regularly in Northern France, in Holland, the extreme south of Sweden, the Baltic islands and the Baltic provinces, thence across Russia to the Urals, southwards to the shores of the Mediterranean, and eastwards to Asia Minor and Palestine. It also breeds in Africa north of the Atlas range.

BREEDING HABITS : The favourite haunts of the Tawny Pipit during the breeding season are sand dunes, dry commons, and wide plains where the soil is loose and sandy. I saw much of the Tawny Pipit in Algeria, especially on the plateaux of the Atlas, where it frequented the rich meadows and barley fields, and notably the wide expanses of fallow land in abundance, where tortoises dwelt in thousands. Although the birds are so common they are not at all gregarious during the breeding season, and live in isolated pairs, but it was no uncommon thing to flush several pairs within a very short distance. During the nesting season the male frequently essays short flights upwards to sing. This bird breeds no earlier in Algeria than in Greece, and probably as late as in Germany. The nest is built amongst the herbage of the plains and fields, sometimes sheltered by a tuft of grass or isolated bush, and sometimes by a projecting stone or earth-clod. It is open and cup-shaped, made externally of dry grass and stalks of plants, straws, and roots, and lined with finer grass and horsehair, although roots are sometimes substituted for the latter in districts where it cannot be obtained. The bird is a close sitter, and when flushed usually flies straight away.

RANGE OF EGG COLOURATION AND MEASUREMENT: The eggs of the Tawny Pipit are five or six in number. They vary in ground colour from pale greenish-blue to white strongly suffused with yellow, mottled, streaked, and spotted with reddish-brown, and with underlying markings of lavender-gray. They are subject to considerable variation. As a rule the markings are numerous, but never so much so as to conceal all the ground colour, and become most dense round the larger end of the egg. Generally the surface spots are large, irregular in shape and pale, but sometimes round and very dark in colour. Average measurement, ·87 inch in length by ·65 inch in breadth. Incubation is performed chiefly by the female, but the duration of the period is apparently unknown.

DIAGNOSTIC CHARACTERS: It is impossible to give any character which will distinguish the eggs of the Tawny Pipit. They can be confused with those of the Crested Lark—which is most unfortunate, as the two species frequent very similar ground—and even with those of the Rufous Warbler, but the latter bird does not breed upon the ground.

Family MOTACILLIDÆ. Genus ANTHUS.
RICHARD'S PIPIT.
ANTHUS RICHARDI, *Vieillot*.
(British : Abnormal autumn Migrant.)
Double Brooded. Laying season, June and July.

BREEDING AREA: East-central Palæarctic region. Richard's Pipit is confined during the breeding season to the great steppe regions of Central Asia. It has been met with breeding as far north as lat. 58° in the valley of the Yenesay, and nests in great numbers on the steppes of the Baikal area, and Dauria, southwards to Eastern Thibet, and thence westwards to Eastern Turkestan.

BREEDING HABITS: The haunts of Richard's Pipit during the breeding season are well-watered plains, wet meadows and pastures, and the swampy areas of the steppes. Like the Meadow Pipit this bird is essentially aquatic in its choice of a breeding place. But little has been recorded of the nidification of this species, owing to observers neglecting their opportunities. It was met with breeding by Dybowsky on the elevated plateaux of the Baikal area, and this naturalist gives many particulars concerning it. Prjevalsky observed it nesting on the East Mongolian steppes, frequenting wet land studded with rushes; and Dr. Scully found it during the summer near Yarkand, frequenting swampy turf-covered ground; whilst lastly, l'Abbé David found it breeding also in Mongolia, and states that it nests amongst grass near water. Incredulous as it may seem, not one of these fortunate naturalists has taken the trouble to describe the nest of Richard's Pipit, and it remains absolutely unknown save to the few men who have taken the eggs. Dybowsky even states that the nest is usually made in a hollow in the ground such as the footprint of a cow or horse. The nest, he says, is very hard to find, the male keeping watch and alarming the sitting female, who leaves the nest and runs along the ground in a Lark-like manner for some distance ere flying up to join her mate, when both endeavour to decoy the intruder away from the spot with anxious notes. The Cuckoo is said by this naturalist usually to select the nest of Richard's Pipit in this locality. There can be little doubt that the nest does not differ in any important respect from those of allied species.

RANGE OF EGG COLOURATION AND MEASUREMENT: The eggs of Richard's Pipit are from four to six in number, five being the usual clutch. They vary in ground colour from pale greenish-white to white suffused with brownish-pink, sprinkled and dusted over the entire surface with olive-brown or reddish-brown, and with indistinct underlying markings of gray. Two fairly distinct types are noticeable, the green-brown spotting being usually correlated with the greenish-white ground; and the red-brown spotting with the brownish-pink ground. Average measurement, ·86 inch in length by ·68 inch in breadth. Incubation is performed by the female, but the duration of the period is unknown.

DIAGNOSTIC CHARACTERS: The size, dusted character of the markings, and the locality of the eggs serve to distinguish them from those of the Rock Pipit, with which perhaps they are most likely to be confused.

Family MOTACILLIDÆ. Genus ANTHUS.

RED-THROATED PIPIT.

ANTHUS CERVINUS (*Pallas*).

(British: Rare abnormal spring migrant.)

Single Brooded. Laying season, June and beginning of July.

BREEDING AREA: Northern Palæarctic region. The Red-throated Pipit breeds on the tundras above the limits of forest growth, from the Atlantic to the Pacific. It breeds locally and in comparatively small numbers in Northern Scandinavia, in Lapland, and Northern Russia. East of the Urals it becomes more abundant, and breeds on the tundras of Siberia as far east as Kamtschatka and, it is to be presumed, Bering Island.

BREEDING HABITS: The Red-throated Pipit is another marsh-loving species, frequenting during the breeding season the vast Arctic tundras, where it is described as being one of the commonest birds. Its favourite haunts are similar to those selected by the Lapland Bunting—swamps where the grass is short, and the wetter portions are divided up into sections by a multitude of drier ridges and tussocks of turf. Although it migrates to its breeding grounds in flocks, the males generally being the first to arrive, the bird is not gregarious during the breeding season, but many scattered pairs dwell in more or less close companionship. This Pipit arrives at its summer haunts shortly after the break-up of winter, and begins to breed very soon after its arrival. Pairing apparently does not take place until the nesting grounds are reached. In its habits it differs little from allied species. The nest is made upon the ground, and the favourite site is in the side of a tussock on the drier part of the tundra. It is cup-shaped, and composed entirely of dry grass, the coarser pieces being used outside and the fine bents for the lining. Of the actions of this Pipit at the nest I find nothing particular recorded.

RANGE OF EGG COLOURATION AND MEASUREMENT: The eggs of the Red-throated Pipit are from four to six in number, and it is interesting to remark that they frequently resemble in their colouration the eggs of the Lapland Bunting, which nests on similar ground. They vary in ground colour from buffish-white to very pale greenish-blue—almost the colour of skimmed milk, spotted and blotched with olive-brown and reddish-brown of various shades, and with underlying markings of paler brown and gray. Two very distinct types occur. On one most of the markings are large and washy, with a few darker specks, but the ground colour is well exposed. On the other the markings are small and uniformly distributed over the entire surface, so that the pale ground is almost entirely concealed. We might even add a third type, in which the markings are very streaky, interspersed with a few spots. Average measurement, ·77 inch in length by ·58 inch in breadth. Incubation is performed by the female, but the duration of the period is unknown.

DIAGNOSTIC CHARACTERS: There is no reliable character by which the eggs of this Pipit may be distinguished from those of the Tree Pipit and those of the Lapland Bunting, but the Tree Pipit does not breed in the Red-throated Pipit's area, and the Lapland Bunting always lines its nest with feathers.

Family CERTHIIDÆ. Genus TICHODROMA.
WALL-CREEPER.
TICHODROMA MURARIA (*Linnæus*).

(British : Very rare abnormal spring and autumn migrant.)
Said to be Double Brooded. Laying season, April to June and July.

BREEDING AREA: Southern Palæarctic and extreme northern Oriental regions. The Wall-Creeper breeds in the mountain districts of Southern Europe and Asia, from the Spanish peninsula to the mountains of Kansu. It breeds throughout the mountain system of Spain and Portugal, in the Pyrenees, the Alps, the Vosges, the mountains of Italy, Sicily, Sardinia, Elba, the Tyrol, and Styria. It also breeds in the Carpathians, the mountains of Greece, Asia Minor, and Palestine, and in the Caucasus. Eastwards it breeds in the mountain ranges of Turkestan, Afghanistan, and Cashmere, and is also an inhabitant of the Himalayas.

BREEDING HABITS: The haunts of the Wall-Creeper are rocky defiles, gorges, and the wall-like cliffs that hem in the mountain torrents. There can be little doubt that this beautiful bird pairs for life, as even in mid-winter it may be seen in company with its mate, and together they perform their wanderings at that season, returning in spring to the old accustomed cliffs where they breed. Although the bird is fairly common in most of the mountains of Southern Europe,—even close to world-famed tourist resorts and notwithstanding the fact of its being such an attractive species, but little has been recorded or observed of its habits by English naturalists. The nest is invariably made in some crevice of the cliffs, often in a situation quite inaccessible. I have examined and described a very perfect nest of this species in the collection of Mr. Seebohm. It is composed principally of moss, amongst which a few grass stalks are interwoven, and the whole is felted together with hair, wool, and feathers. The lining is also composed of wool and hair massed very compactly together. It is an open cup-shaped structure, about six inches in diameter over all, the cavity containing the eggs being about three inches across and one and a half inches deep. Of the actions of the birds at the nest all observers appear to be silent.

RANGE OF EGG COLOURATION AND MEASUREMENT : The eggs of the Wall-Creeper are from three to five in number. They are pure white in ground colour, minutely freckled with reddish-brown, and with numerous underlying markings of lilac-gray. As a rule most of the spots are on the large end of the egg. Average measurement, ·78 inch in length by ·56 inch in breadth. It is not known whether male or female, or both, incubate the eggs, and the duration of the period remains undetermined.

DIAGNOSTIC CHARACTERS : The size and minute markings of the eggs of the Wall-Creeper distinguish them from those of the Common Creeper, but from eggs of the Wren and the Nuthatch they are not so readily separated ; the style and situation of the nest is, however, a sufficient guide to their correct identification.

Family PARIDÆ. Genus REGULUS.
Sub-family REGULINÆ.

FIRECREST.

REGULUS IGNICAPILLUS (*Brehm*).

(British : Abnormal autumn and winter migrant.)

Single Brooded. Laying season, March to May, according to latitude.

BREEDING AREA : South-west Palæarctic region. The Firecrest is somewhat restricted in its distribution during the breeding season. It breeds locally in the Baltic Provinces (although, according to Herr E. Hartert, there is no record of its occurrence in East Prussia), in Central and Southern Germany, in France, the Spanish Peninsula, various islands in the Mediterranean, Italy, Switzerland, Austria, Hungary, the Danubian Provinces, Turkey, Greece, Southern Russia, and Asia Minor. South of the Mediterranean it breeds commonly in the mountain districts of the Atlas.

BREEDING HABITS : The Firecrest resembles the Goldcrest very closely in its habits, and to a great extent the haunts of the two species are similar. The favourite breeding places of the Firecrest in Europe are woods and plantations of firs, even small clumps of these trees in gardens and pleasure-grounds being frequented. In Algeria, where I met with this species in abundance, its favourite haunts are cedar forests, and the extensive evergreen oak woods that clothe the sides of the mountains. The Firecrest most probably pairs for life, and although a new nest is made each spring, the birds resort to certain trees with much attachment. The bird is sociable enough in autumn and winter, but always breeds in scattered pairs. The nest is like that

E

of the Goldcrest, nearly globular in shape, and slung from the drooping extremities of a branch, several of the twigs being interwoven with the structure. It is composed of moss, bound together with spiders' webs, and studded with lichens, and warmly lined with great quantities of feathers. In Algeria, and other places, where the timber is heavily draped with long lichens and tree moss, this material forms the greater part of the nest, and renders its discovery most difficult. In the vicinity of the nest the Firecrest is remarkably shy and retiring, and the female sits closely.

RANGE OF EGG COLOURATION AND MEASUREMENT: The eggs of the Firecrest are from six to ten in number, seven or eight being a usual clutch. They are very pale brick-red in ground colour, mottled, clouded, and speckled over the whole surface with brownish-red. Occasionally a few specks and short streaks of darker brown occur. As a rule the markings are evenly distributed and conceal most of the ground colour, but on some varieties the surface colour is most abundant at the large end of the egg, in the form of a cap or zone. Average measurement, ·53 inch in length by ·43 inch in breadth. Incubation, performed chiefly by the female, lasts about fourteen days.

DIAGNOSTIC CHARACTERS: The only egg with which the egg of the Firecrest can be confused is that of the Goldcrest, but it is readily distinguished by its much redder appearance.

Family AMPELIDÆ. Genus AMPELIS.

WAXWING.

AMPELIS GARRULUS, *Linnæus*.

(British : Nomadic autumn and winter migrant.)

Single Brooded. Laying season, June.

BREEDING AREA : Northern Nearctic and Palæarctic regions. The Waxwing is very erratic in its choice of breeding grounds, changing them more or less capriciously from year to year. It breeds in the pine forests near the Arctic circle in both hemispheres, but its exact nesting places are little known. It is widely distributed, if local, in Lapland and Finland ; it has been met with during summer in the valleys of the Petchora and Yenesay. On the American continent its eggs have been taken near Fort Yukon in Alaska, and the bird observed during the breeding season in the valley of the Anderson river, which locality may possibly mark the limit of its eastern distribution in the New World.

BREEDING HABITS : The nidification of few birds is surrounded with such a voluminous literature as that of the Waxwing, but unfortunately most of it is the veriest padding, and contains little practical information. The discovery of the nest and eggs of the Waxwing is due to the unwearied exertions of the late John Wolley, who spent no less than five consecutive summers and two of the four intervening winters in Lapland in eager quest of them. Previous to his discoveries the nidification of the Waxwing was surrounded by romance, and the eggs were generally presumed to be laid in holes in trees and rocks ! The Waxwing, like the Rose-coloured Starling, is very erratic in its choice of a breeding place, and appears to settle in the nearest convenient spot to where its nomadic winter wanderings have led it. One season it may breed in a locality in vast abundance, and perhaps not a solitary pair will resort to the old station the season following. The Waxwing is chiefly gregarious during the nesting period, and breeds in large scattered colonies. Of its pairing habits nothing appears to have been observed. Its favourite breeding grounds are the more open forests of fir and spruce intermixed with birches. The nest is made at a moderate height from the ground— eight to twelve feet—on a branch, and is composed externally of dead twigs and reindeer-moss, and lined with dry grass, quantities of a hair-like black tree lichen, strips of inner birch bark, and feathers. It is cup-shaped, bulky, and rather deep, the cavity containing the eggs being about four inches in diameter and two inches in depth. Many nests of the Waxwing have been secured since Wolley's day, but the eggs have never been taken in such vast numbers as they were by his collectors in the summer of 1858, when the spoil reached the tempting total of nearly seven hundred eggs from nearly one hundred and fifty nests!

RANGE OF EGG COLOURATION AND MEASUREMENT: The eggs of the Waxwing are from five to seven in number. Most of the magnificent series of eggs obtained by Wolley remain in the collection of Professor Newton, and I cannot do better than give the description of them in the latter gentleman's own words. " The ground is most generally of a delicate sea-green, sometimes fading to French white, but very often of a more or less pale olive, and occasionally of a dull purplish-gray. On this are almost always bold blotches, spots, and specks of deep brownish-black [blackish-brown would be a better expression, as no eggs are known to be marked with *black*], though sometimes the edges are blurred. Beneath these stronger markings there is nearly always a series of blotches or streaks of grayish-lilac, and among them well-defined spots or specks of yellowish-brown are interspersed. In some eggs the darkest markings are quite wanting, in others the ground is of a deep olive colour." Average measurement, ·97 inch in length by ·68 inch in breadth. Incubation appears to be performed by both sexes, but the duration of the period is unknown.

DIAGNOSTIC CHARACTERS: The eggs of the Waxwing cannot readily be confused with those of any other species, except with those of the American Waxwing or Cedar Bird, and possibly with those of the Japanese Waxwing. They are, however, normally much larger; and the locality is of some importance in determining them.

Family LANIIDÆ.   Genus LANIUS.

LESSER GRAY SHRIKE.

LANIUS MINOR, *Gmelin*.

(British : Rare abnormal spring and autumn migrant.)

Single Brooded. Laying season, latter half of May and in June.

BREEDING AREA: South-west Palæarctic region. The Lesser Gray Shrike breeds in the Baltic Provinces of Prussia, in Germany, France (as far west as the valley of the Rhone), Switzerland, Italy, Sicily, Dalmatia, Austria, the Danubian Provinces, Turkey, Greece, Russia (south of lat. 57°), Asia Minor, Palestine, Persia, Turkestan, and Siberia as far north as Omsk in the valley of the Irtish, and as far east as Lake Saisan.

BREEDING HABITS: The Lesser Gray Shrike is a late migrant to Europe, not reaching its breeding grounds even in the south before the middle of April and in Germany not before the beginning of May. The favourite breeding haunts of this Shrike are in cultivated districts, in the vicinity of gardens and plantations; in some localities it prefers the shrub-covered sides of rocky slopes, amongst which are scattered small trees. The bird makes itself very conspicuous, sitting on the tops of bushes and trees. It appears to pair for life, and will yearly return to a favourite nesting place. The nest is generally made in a fruit or olive tree, a poplar, an oak, or a mulberry, usually from ten to fifteen feet from the ground. It is a large, bulky, cup-shaped structure, composed externally of roots, twigs, coarse grass, straws, and stalks of plants, and lined with wool, hair, feathers, and some aromatic plant, such as lavender or thyme. The nests vary a good deal in materials. Some are made entirely of an aromatic plant, others almost completely of cudweed; whilst in Prussia Herr E. Hartert states that the nest is invariably lined with these strongly-scented plants. If the nest is menaced by predaceous birds or animals the parents become very bold and pugnacious.

RANGE OF EGG COLOURATION AND MEASUREMENT: The eggs of the Lesser Gray Shrike are from four to seven in number, and vary considerably in size and shape. They are pale bluish-green in ground colour, spotted and blotched with olive-brown, and with underlying markings of pale greenish-brown. The spots normally are somewhat large and irregular, and in exceptionally fine examples the markings form a zone round the larger end, the spaces between the larger masses of surface colour being filled in with the smaller underlying spots. Some eggs are very sparsely spotted, chiefly on the larger end. The rufous type of egg does not appear to occur in this species. Average measurement, ·98 inch in length by ·77 inch in breadth. Incubation, performed almost entirely by the female, lasts from fifteen to sixteen days.

DIAGNOSTIC CHARACTERS: The green ground colour of the eggs of the Lesser Gray Shrike readily distinguishes them from those of allied species, more especially when combined with size.

Family LANIIDÆ.   Genus LANIUS.

GREAT GRAY SHRIKE.

LANIUS EXCUBITOR, *Linnæus*.

(British: Uncommon autumn migrant.)

Single Brooded. Laying season, latter half of May and in June.

BREEDING AREA: Western Palæarctic region. The Great Gray Shrike breeds from about lat. 70° in Scandinavia, and in Russia as far east as the Urals, southwards through Denmark, Holland, Belgium, Germany, France, and various parts of Central Europe. Owing to interbreeding with sub-specific forms, it is difficult to define the exact area occupied by this species.

BREEDING HABITS: The Great Gray Shrike is very partial to exposed haunts during the non-breeding season, and renders itself conspicuous enough, like all birds of its kind, by sitting on the topmost twigs of bushes and low trees; but when the nesting period arrives it shows a preference for more sheltered localities. Its favourite breeding haunts are in plantations and the open places in woods. Nothing appears to be known respecting the pairing habits of this species, or whether it is in the habit of returning yearly to a particular spot to breed, as the Red-backed Shrike invariably does. The nest is placed in trees, either evergreens or deciduous, usually in a fork of a large tree or at the summit of a small one. It is large, bulky, and cup-shaped, composed externally of dead twigs, dry grass, roots, and stalks of plants, lined with finer roots, wool, hair, and feathers, the lining material depending a good deal on what may chance to be available, but always of some soft substance. This Shrike is a close sitter, even before incubation has actually commenced, and when disturbed frequently flies about from tree to tree in the vicinity, and shows much solicitude for its nest. It is bold enough in driving off predaceous birds from the spot.

RANGE OF EGG COLOURATION AND MEASUREMENT: The eggs of the Great Gray Shrike are from five to seven in number. They vary from buffish-white to the palest of green in ground colour, spotted, blotched, and speckled with olive-brown, and with underlying markings of lilac-gray. As a rule most of the surface spots, which vary from light to dark brown, are on the largest end of the egg, are seldom very clearly defined, but many are often confluent, and frequently form a zone. I have never seen the red type of egg in this species. Average measurement, 1·1 inch in length by ·8 inch in breadth. Incubation, performed almost entirely by the female, lasts fifteen or sixteen days.

DIAGNOSTIC CHARACTERS: The eggs of the Great Gray Shrike are generally easily distinguished by their size, but in all cases I would advise careful identification at the nest. The nest and eggs might easily in some districts be confused with those of *Lanius leucopterus*, and even more so with *L. excubitor major*.

Family TURDIDÆ.   Genus PHYLLOSCOPUS.
Sub-family SYLVIINÆ.

YELLOW-BROWED WILLOW WREN.

PHYLLOSCOPUS SUPERCILIOSUS (*Gmelin*).

(British : Very rare abnormal autumn migrant.)

Single Brooded.   Laying season, latter end of June.

BREEDING AREA : North-east Palæarctic region. The Yellow-browed Willow Wren is presumed to breed in the pine regions of Siberia, from the valley of the Yenesay eastwards to the Pacific as far north as the Arctic circle, and as far south as the mountains in the Baikal area. It has, however, only been discovered breeding in the valley of the Yenesay, where its nest and eggs were taken by Mr. Seebohm during the summer of 1877.

BREEDING HABITS : But little is known of the nidification of the Yellow-browed Willow Wren. The account of the nesting of this species given in Professor Newton's edition of Yarrell's *History of British Birds*, and in Mr. Dresser's *Birds of Europe*, refer to a totally distinct species, an Indian Willow Wren (*Phylloscopus humii*). To Mr. Seebohm we are entirely indebted for a description of the breeding habits of this interesting little bird. It did not arrive at its breeding haunts until the fourth of June, when, in company with the Common Willow Wren and the Siberian Chiffchaff, it was observed amongst the bare branches of the willow and birch trees on the banks of the Yenesay, where the snow had melted. This species evidently migrates in parties, and probably pairs after its arrival. Whether the sexes separate to perform the journey, as the Common Willow Wren does, appears not to be known. Probably such is the case, as Mr. Seebohm informs us that the birds were in no apparent hurry to breed, and the males were possibly awaiting the arrival of their partners. About a fortnight after its arrival nest-building commenced. The favourite haunts were the pine forests on the banks of the Koo-ray-i-ka and the Yenesay. The song is apparently little more than that of the Wood Wren, and uttered with the same shivering of the wings. The nest found by Mr. Seebohm, and which together with the eggs I have had the pleasure of examining, was made amongst the bilberry wires, moss, and grass at the foot of a birch tree. It is semi-domed, like that of the Willow Wren, but more open than that of the Chiffchaff, and made externally of dry grass and bits of moss, and lined with reindeer-hair. The actions of the bird at the nest are precisely the same as those of the British species. It is very restless, hopping about in the vicinity of its home, and tiring out all but the most persevering patience, before betraying its secret.

RANGE OF EGG COLOURATION AND MEASUREMENT: The eggs of the Yellow-browed Willow Wren, so far as is known, are six in number. They are pure white in ground colour, spotted and speckled with reddish-brown, and with a few underlying markings of paler brown. The spots, which are numerous, are mostly on the larger end of the egg, some of them confluent and forming an irregular zone. Average measurement, ·6 inch in length by ·45 inch in breadth. The period of incubation, of course, remains unknown, and whether one or both parents perform the task was unfortunately not remarked.

DIAGNOSTIC CHARACTERS: The eggs of this species will always require the most careful identification. Indeed but few ornithologists, in my opinion, are competent to take them at all, and I know of no character which will distinguish them from those of several allied species.

Family TURDIDÆ. Genus SYLVIA.
Sub-family SYLVIINÆ.

ORPHEAN WARBLER.

SYLVIA ORPHEA, *Temminck*.

(British: Possibly breeds; very rare abnormal spring migrant.)
Single Brooded. Laying season, end of April and in May.

BREEDING AREA: South-west Palæarctic region. The Orphean Warbler breeds in Central and Southern France, German Lorraine, throughout the Spanish Peninsula in suitable districts, in Dalmatia, Turkey, Greece, Southern Russia, Asia Minor, and Palestine. It should be remarked that examples from the two latter countries are intermediate between typical *Sylvia orphea* and *S. orphea jerdoni*, which ranges through Persia and Turkestan to India. South of the Mediterranean the typical Orphean Warbler breeds in Morocco and Algeria.

BREEDING HABITS: As is almost universally the case with birds that do not breed so far north or west as the British Islands, the Orphean Warbler is a somewhat late migrant, not reaching its more southerly European haunts before the first or second week in April. Like its near ally the Blackcap, it is a secretive species, and loves to frequent localities in which there is plenty of cover. Olive groves, cork woods, and vineyards are a favourite resort, as well as the rough uncultivated ground covered with bushes and thickets, between the zone of the vines and olives and the pine zone. Of the pairing habits of this species nothing appears to be recorded. It probably pairs annually. The nest is made in a bush of some kind or in a low tree—it has been found in a cork-oak twelve feet from the ground. It is a fairly compact structure, cup-shaped, composed externally of dry grass and the fine stalks of plants, and lined with finer grass and a small quantity of vegetable down. It is somewhat remarkable that such a lining should be added by a species breeding in such a high temperature, while the Blackcap, which so frequently nests in a much lower temperature, makes no such provision. Of the actions of this bird at the nest nothing is recorded.

RANGE OF EGG COLOURATION AND MEASUREMENT: The eggs of the Orphean Warbler are four or five in number, generally the latter. They are white, or tinged with gray or brown in ground colour, spotted and blotched with olive-brown and very dark brown, and with underlying markings of pale gray or pale brown. Most of the markings are on the larger end of the egg, and often form an irregular zone; the smaller spots are generally most intense in colour, and many of the larger ones are often confluent and run into irregular blotches or streaks. Average measurement, ·8 inch in length by ·6 inch in breadth. Incubation, performed chiefly by the female, lasts fourteen days.

DIAGNOSTIC CHARACTERS: The size and general colouration of the eggs and the structure of the nest is a tolerably safe guide to the determination of the eggs of the Orphean Warbler, but they should always be carefully authenticated. The Cuckoo frequently uses the nest of this species, and usually deposits an egg so closely resembling those of the Orphean Warbler in appearance that it is only determined by microscopic examination. It is, however, always larger.

Family TURDIDÆ. Genus SYLVIA.
Sub-family SYLVIINÆ.

BARRED WARBLER.

SYLVIA NISORIA (*Bechstein*).

(British: Very rare abnormal autumn migrant.)

Single Brooded. Laying season, latter end of May and in June.

BREEDING AREA: South-western Palæarctic region. The Barred Warbler breeds sparingly in South Sweden and Denmark, more commonly in Germany east of the valley of the Rhine, Northern Italy, Transylvania, Bulgaria, Turkey, Southern Russia, Persia, and Turkestan up to an elevation of from 6,000 to 10,000 feet.

BREEDING HABITS: The Barred Warbler is another late migrant, not reaching its breeding grounds before the very end of April or early in May. Nothing appears to be known respecting its pairing habits, but the bird probably mates annually. Its favourite resorts during the nesting season are small plantations and ground covered with scrub and thickets. Owing to its shy, retiring disposition it is much liable to be overlooked. The nest is generally placed in a thick bush a few feet above the ground, but instances are on record where it has been found almost on the ground itself, whilst, in one instance only, it has been discovered at the summit of a birch tree, twenty-five feet from the ground. It is a somewhat bulky structure, and though net-like and flimsy-looking, is rather more compactly made than is usual in this class of birds. It is cup-shaped, and made externally of dry stalks and roots, small withered plants, and occasional scraps of thistledown or cocoons of insects, and is lined with finer roots and horsehair. The bird is a close sitter, but not demonstrative, and when flushed conceals itself amongst the surrounding vegetation, the male often uttering his warning *tec* or *rar*.

RANGE OF EGG COLOURATION AND MEASUREMENT: The eggs of the Barred Warbler are from four to six in number, five being the average clutch. They vary from grayish-white to buffish-white in ground colour, mottled and freckled with gray, and more rarely spotted with brown. The eggs of this bird are very remarkable, inasmuch that most of the markings are underlying ones, or covered with a thin layer of ground colour, the surface spots being few and indistinct. A rare type, however, occurs in which these conditions are reversed, and the surface spots are larger, clearly defined, and greatly outnumber the underlying markings. Herr E. Hartert records a very handsome clutch of the latter type, in which the eggs are "spotted with deep chestnut-brown." Average measurement, ·85 inch in length by ·62 inch in breadth. Whether both sexes incubate, or only one, appears not to be known, as is also the duration of the period.

DIAGNOSTIC CHARACTERS: The eggs of the Barred Warbler are very distinct, and cannot readily be confused with those of any other species. The predominating underlying markings, combined with the size of the egg, are safe distinguishing characters.

Family TURDIDÆ.   Genus AËDON.
Sub-family SYLVIINÆ.

RUFOUS WARBLER.

AËDON GALACTODES (*Temminck*).

(British : Very rare abnormal autumn migrant.)

Single Brooded. Laying season, latter half of May and in June.

BREEDING AREA : Extreme south-western Palæarctic region. The Rufous Warbler breeds commonly in the Spanish Peninsula, in Morocco, Algeria, Tunis, Tripoli, Egypt, and Palestine as far north as about lat. 34°.

BREEDING HABITS: The Rufous Warbler is another very late migrant, not crossing the Mediterranean before the end of April. In its choice of a haunt the Rufous Warbler shows considerable divergence from that of its near relations, often frequenting very bare and sterile spots. In Algeria I remarked its preference for the Arab gardens of prickly pears, and the date palm forests where little undergrowth is to be seen. In other localities it selects olive groves and vineyards, and the thickets of tamarisk by the river sides. This Warbler pairs annually, and in the season of courtship the male may often be seen in chase of the female, both spending much of their time upon the ground. The nest is placed in a tamarisk tree or bush, in a hedge, or between the leaves of a cactus, at heights varying from one to six feet from the ground. Mr. Salvin records a nest which he found amongst the exposed roots of a tree on a bank. It is a large and somewhat bulky, cup-shaped structure, composed externally of dead shoots, roots, straws, coarse grass, and bits of lichen, lined with wool, vegetable down, a few feathers or hair. The lining varies a good deal according to locality, but is always composed of some soft substance, and almost invariably includes a piece of dry serpent's skin—a charm to protect the eggs from snakes, according to popular legend. The birds make little or no attempt to conceal their nest, which from its very conspicuousness often escapes discovery.

RANGE OF EGG COLOURATION AND MEASUREMENT: The eggs of the Rufous Warbler are from three to five in number, the latter being the usual clutch. They vary from pale gray to very pale blue in ground colour, spotted with pale and dark brown, and with underlying markings of lilac-gray. Two distinct types occur. The first has the markings large and splashed, and sparingly streaked; the second has them finely sprinkled over the entire surface, but most numerous at the larger end of the egg. As a rule the first type is correlated with the gray and the second with the pale blue ground colour, and in the first the underlying markings are the most apparent. Average measurement, ·87 inch in length by ·63 inch in breadth. Incubation is performed chiefly by the female, but the duration of the period is unknown.

DIAGNOSTIC CHARACTERS: The eggs of the Rufous Warbler are fairly distinct, their size and colouration preventing confusion with those of allied species. From the eggs of the Tawny Pipit they are not so readily distinguished, but always appear to be less bulky. Careful identification is required, although the nesting habits of the two birds are different.

Family TURDIDÆ. Genus ACROCEPHALUS.
Sub-family SYLVIINÆ.

GREAT REED WARBLER.
ACROCEPHALUS TURDOIDES (*Meyer*).
(British : Very rare abnormal spring and autumn migrant.)
Single Brooded. Laying season, latter end of May and early June.

BREEDING AREA : South-western Palæarctic region. The Great Reed Warbler breeds in all suitable districts throughout Europe south of the Baltic and the British Islands. South of the Mediterranean it is said to breed in Morocco and Algeria ; whilst eastwards it breeds in Palestine, Asia Minor, Northern Persia, and Turkestan.

BREEDING HABITS : It is rather a remarkable fact that the Great Reed Warbler occurs so rarely in the British Islands, seeing that it breeds commonly in Holland, Belgium, and even so close as Calais, almost within sight of our shores. As is usual with such species, it is a remarkably late migrant, not arriving at its most northerly breeding places before the first or second week in May. The sole breeding haunts of this fine Warbler are the belts and forests and beds of the common reed (*Arundo phragmites*). It is a rather local species, but frequents reeds on small ponds as well as those that fringe lakes, broads, and slow running rivers. It probably pairs annually, but our information on this point is scanty. The nest is almost invariably supported by from three to five stems of the reeds, and is situated about midway from the water to their summits. The selected reeds are generally well in the thicket where sufficient seclusion is afforded. Instances are, however, on record where the nest of this species has

F

been found on the twigs of willows (*fide* Hartert), and it has been known, according to the same authority, "high up in a birch tree." The nest is large and strongly put together, funnel-shaped and deep, and the conical base which supports the cup of the nest is often bulky. Externally the nest is made of the dry withered leaves of the reeds, intermixed with a few roots and withered reed flowers, and lined with the latter together with a few grass stems. Sometimes one or two leaves growing on the selected reeds are interwoven with the nest, and some nests are lined with scraps of moss, vegetable down, or even a few feathers. Externally it is about five inches in height, but not more than half this space is occupied by the cavity containing the eggs. When disturbed from the nest the parent birds evince considerable anxiety, venturing close to the intruder, and resenting interference with their home by uttering croaking notes of displeasure.

RANGE OF EGG COLOURATION AND MEASUREMENT: The eggs of the Great Reed Warbler are from four to six in number. They vary from pale greenish-blue to blue more strongly tinged with pale green or gray, blotched and spotted with olive-brown and russet-brown, and with underlying markings of gray. The markings are generally very handsome and bold blotches, intermixed with smaller spots, and the gray underlying markings are pretty evenly distributed over the entire surface. The eggs vary much in size: average measurement, ·9 inch in length by ·65 inch in breadth. Incubation, performed chiefly by the female, lasts, according to Thienemann, fourteen or fifteen days.

DIAGNOSTIC CHARACTERS: The eggs of the Great Reed Warbler are readily distinguished from those of allied species by their much larger size.

Family TURDIDÆ. Genus ACROCEPHALUS.
Sub-family SYLVIINÆ.

AQUATIC WARBLER.

ACROCEPHALUS AQUATICUS (*Gmelin*).

(British : Very rare abnormal autumn migrant.)

Single Brooded. Laying season, latter half of May and early June.

BREEDING AREA: South-western Palæarctic region. The Aquatic Warbler has a somewhat restricted breeding range, being for the most part confined to Central Europe. It breeds in Denmark, Germany, the Netherlands, France, Italy (including Sardinia and Sicily), Austria, and Central Russia, as far east as the Southern Urals. South of the Mediterranean it is said to breed sparingly in Algeria and Tunis, but its distribution is very imperfectly defined.

BREEDING HABITS: The habits of the Aquatic Warbler are very similar to those of the Sedge Warbler, and the kind of localities it frequents are very much the same. It reaches its more northern summer haunts during the third week in April. Its favourite haunts are swampy localities, the vegetation on the banks of rivers and ponds ; ditches which are almost choked with aquatic herbage, brambles, and briars ; thickets of willows and osier beds. It is a shy and secretive species, keeping much to the cover of the vegetation. It pairs soon after its arrival, but very often nesting does not commence for several weeks after that event. The nest is open and cup-shaped, and though never exactly upon the ground is often built a few inches above it, and rarely more than a foot or eighteen inches from it. It is said to be suspended between the stalks of plants growing round it, and which are often interwoven with it, in the tussocks of sedge or the small thorns and willows by the water side. It is a small and carelessly made structure, but the cup is neat and rounded, composed externally of dry coarse grass and roots, and lined with fine grass and horsehair. Occasionally cocoons, feathers, and the flowers of aquatic plants are woven into the nest, but the lining appears always to be hair. The bird is a close sitter, very skulking when disturbed, and fond of manifesting its displeasure at the intrusion by a series of scolding cries.

RANGE OF EGG COLOURATION AND MEASUREMENT : The eggs of the Aquatic Warbler are from four to six in number, five being a usual clutch. They are brownish-white in ground colour, clouded and mottled over the whole surface with yellowish-brown, and occasionally streaked with a few hair-like lines or scrolls of very dark brown. Average measurement, ·68 inch in length by ·51 inch in breadth. Incubation, performed chiefly by the female, lasts fourteen or fifteen days.

DIAGNOSTIC CHARACTERS : Unfortunately there is no character by which the eggs of the Aquatic Warbler may be distinguished from those of the Sedge Warbler. Eggs that are not thoroughly and trustworthily identified are not of the slightest value.

Family TURDIDÆ. Genus HYPOLAIS.
Sub-family SYLVIINÆ.

ICTERINE WARBLER.

HYPOLAIS HYPOLAIS (*Linnæus*).

(British : Very rare abnormal spring and autumn migrant.)

Single Brooded. Laying season, latter end of May and in June.

BREEDING AREA: West-central Palæarctic region. The Icterine Warbler breeds in Norway as far north as the Arctic circle, but in Sweden and West Russia not beyond lat. 65°, and in East Russia no higher than lat. 57°. Southwards it breeds in Central Russia, the Baltic Provinces, throughout Germany, Denmark, Holland, Belgium, Northern France, Italy, and Sicily.

BREEDING HABITS : The Icterine Warbler is another very late migrant, not reaching its more northerly breeding grounds before the first or even the second week in May. It is probable that the males arrive a little before the females, and song does not commence until the latter appear. Pairing takes place annually. The favourite breeding places of this Warbler are gardens and hedgerows which contain trees and high bushes, orchards and well-cultivated lands near to houses. During the nesting season the male is a very persistent singer, very jealous of his particular haunt, and ready to quarrel with and beat off any intruder. The nest is usually made in a fork of the branches of a small tree or large bush—a lilac tree is a favourite spot—not more than eight or ten feet from the ground. It is made on a somewhat similar plan to that of the Goldfinch, but in a series of nests great variation is to be observed, some being much better made than others. It is cup-shaped, and made externally of moss and dry grass, strips of bark and roots, felted together and cemented with cocoons, thistle-down, wool, and flakes of lichen, and lined with fine round grass stems and horsehair. The bird is a close sitter, and when disturbed glides up and down the cover in a restless manner, uttering a shrill *tec* of alarm and remonstrance, often in company with its mate.

RANGE OF EGG COLOURATION AND MEASUREMENT: The eggs of the Icterine Warbler are from four to six in number, five being the usual clutch. They vary in ground colour from grayish-pink to pink with a brownish cast, spotted over the entire surface, and occasionally streaked with dark blackish-brown, and with indistinct underlying markings of paler brown. One type has numerous fine streaks of paler brown; another has the spots small and dusted over the surface. Average measurement, ·72 inch in length, by ·55 inch in breadth. Incubation, performed chiefly by the female, lasts fourteen days.

DIAGNOSTIC CHARACTERS: The rose-pink ground colour is a reliable character to distinguish the eggs of the Icterine Warbler from those of most other species. I do not, however, find any reliable character to distinguish them from those of the nearly allied *Hypolais polyglotta*. They require careful identification and reliable authentication.

Family TURDIDÆ. Genus GEOCICHLA.
Sub-family TURDINÆ.

WHITE'S GROUND THRUSH.

GEOCICHLA VARIA (*Pallas*).

(British : Very rare abnormal autumn migrant.)
Number of broods unknown. Laying season, May (?).

BREEDING AREA: Eastern Palæarctic region. White's Ground Thrush is presumed to breed in South Siberia from the Baikal area eastwards along the valleys of the Amoor, in North-east Mongolia, Manchooria, and Northern China, but its alleged nest has only been taken in the latter locality. Dr. Menzbier is of the opinion that this Thrush is distributed during the breeding season "throughout the whole wooded districts of Siberia" to as far west as the forests of the Urals. In support of this latter statement he records (*Ibis*, 1893, p. 372) three examples obtained in July and August in the Governments of Ufa and Perm. Prof. Kovtzov states that it is rare in Southern Tobolsk and common in Northern Admalinsk.

BREEDING HABITS : Our information respecting the nidification of White's Ground Thrush is only of the most meagre kind, and to a certain extent of a very unsatisfactory nature. Indeed, cautious naturalists are fully justified in casting doubt upon the discovery of the nest and eggs of this species, inasmuch that they were never authenticated in the only possible way that could dispel all uncertainty, namely, the shooting of one or both parents at the nest. The late Mr. Swinhoe, who obtained this nest and eggs in Northern China in the spring of 1872, is known to have been a careful student ; but as he only saw the birds that he presumed rightly or wrongly to belong to the nest which some boys had taken from a tree just previously, there must always remain an element of doubt until birds are absolutely shot at or seen on the nest. This nest was made in the upper branches of a pine tree. From Mr. Swinhoe's collection it passed into that of Mr. Seebohm, where I had the opportunity of examining it, together with the eggs. It is about the size of that of the Blackbird, and made on a very similar plan. Externally it is composed of fine and coarse grass, dead rush leaves, moss, a few twigs, and an occasional withered leaf; in this shell a first lining of wet mud had been placed, amongst which are sticking a few bits of weed, which were doubtless growing in the mud when it was taken to the nest; the final lining consists of coarse fibrous roots and a few bits of sedge. The birds are described as being very anxious when their nest was removed, and were recognized by Mr. Swinhoe as the present species.

RANGE OF EGG COLOURATION AND MEASUREMENT: The nest obtained by Mr. Swinhoe contained three eggs, but probably this number does not represent a full clutch. They are greenish-white in ground colour, minutely and evenly spotted with reddish-brown over most of the surface. Average measurement, 1·2 inch in length by ·9 inch in breadth. The duration of the incubation period is unknown.

DIAGNOSTIC CHARACTERS: As the eggs of this species may possibly be yet unknown to science, it is not wise to form any diagnosis from the above description, even if such were possible. There is always the possibility that this nest and eggs belonged to *Merula mandarina*, a species to which Mr. Swinhoe himself at first thought them to belong, and to which admittedly they bear a very strong resemblance.

Family TURDIDÆ. Genus TURDUS.
Sub-family TURDINÆ.

FIELDFARE.

TURDUS PILARIS, *Linnæus.*

(British : Common autumn migrant.)

Single Brooded probably. Laying season, May to beginning of July.

BREEDING AREA: Northern Palæarctic region. The Fieldfare breeds commonly throughout Scandinavia, Finland, Northern Russia, and Siberia, as far east as the watershed of the Yenesay and Lena. Southwards it breeds more sparingly and locally in Central Russia, the Baltic Provinces, Prussia, Poland, and Central Germany, in Bavaria, and the Austrian Provinces of Bohemia and Moravia. That it still continues to discard the British Islands as a breeding place is very remarkable, but is probably due to the much higher spring temperature prevailing in them.

BREEDING AREA: The Fieldfare reaches its breeding places in May or June according to latitude. Its favourite nesting grounds are the more open pine, fir, and birch woods, but in more northern regions, where the timber becomes small and stunted, the thickets of willows and birches afford a haunt; whilst on the Arctic tundras the bird is compelled to nest upon or near the ground like the Ring Ouzel on our own moors. The Fieldfare may pair annually, but of this we know nothing; I suspect that it may possibly pair for life and return each season to old breeding colonies, but doubtless the old nests are renewed each year. During the laying season the male is most musical, often commencing to sing whilst flying to a perching place. The bird for the most part is gregarious during the breeding season, and nests in colonies of varying size according to the abundance or rarity of the individuals in a certain area. Many outlying nests are made some distance from each other on the outskirts of the colony, but in the colony itself they are close together. The nests are made on the flat branches of the pines, in forks of the birch trees close to the trunks, or in suitable crotches in the alders. Occasionally an odd nest will be made in an outhouse or amongst a heap of wood near the peasants' cottages. Nests on the tundras are usually placed on the ground near a ridge or ledge. Mr. Seebohm remarked that the colonies of this species were not so large in Siberia as in Norway, and that the bird either bred in isolated pairs or in small parties. The nest is made on a similar plan to that of the Blackbird, being composed externally of dry grass, moss, and a few twigs, then lined with mud, and finally with an abundance of fine grass. The birds become very noisy when disturbed, but do not remain long in the vicinity of their nests, being nothing near so bold as the Missel Thrush. The birds in a colony do not all begin nesting together, and young birds and eggs in various stages of development may be found at the same time, just as is the case in a rookery or a gullery.

RANGE OF EGG COLOURATION AND MEASUREMENT: The eggs of the Fieldfare are usually from four to six in number, but occasionally seven are found, and exceptionally only three. They are subject to enormous variation, but probably every type is represented in the eggs of the Blackbird and the Ring Ouzel. They vary in ground colour from bluish-green to greenish-blue, blotched, spotted, freckled and marbled over the entire surface with rich reddish-brown. A rare variety is almost blue in ground colour with only a few rich brown streaks or spots chiefly at the larger end. Generally the markings are most numerous and intricate over the larger end of the egg and conceal much of the ground colour, but varieties are not uncommon in which the blotches are scattered, and show much of the ground colour between them. A few pale gray underlying markings are visible on many eggs. Average measurement, 1·2 inch in length by ·85 inch in breadth. Incubation, performed by both sexes, lasts from fourteen to sixteen days.

DIAGNOSTIC CHARACTERS: It is impossible to give any character by which the eggs of the Fieldfare can be distinguished from those of the Blackbird or the Ring Ouzel. Both the latter species are non-gregarious, however; from eggs of the Redwing they are easily separated by their larger size and generally handsome appearance. They should be carefully identified, however, in every case.

Family TURDIDÆ.
Sub-family *TURDINÆ*.

Genus TURDUS.

REDWING.

TURDUS ILIACUS, *Linnæus*.

(British : Common autumn migrant.)

Single Brooded. Laying season, May and June to middle of July.

BREEDING AREA: Northern Palæarctic region. The Redwing breeds in suitable localities throughout Scandinavia, and in Russia from about lat. 54° northwards to the coast. Herr E. Hartert states that it breeds regularly near Memel in the extreme north-east of Prussia, and that it occasionally nests in Germany. East of the Urals the Redwing breeds in Northern Siberia, probably from about lat. 60°, and extending north beyond the limits of forest growth to at least lat. 71°, but becomes very rare east of the valley of the Yenesay.

BREEDING HABITS: The Redwing arrives at its breeding grounds in Scandinavia towards the end of April or early in May. Further east in Northern Russia where the springs are later it seldom arrives before the middle of May; whilst in Siberia it does not make its appearance until the first week in June. The breeding haunts of the Redwing are very similar to those of the Fieldfare, indeed odd pairs of the present species may frequently be found nesting within the limits of that bird's colonies. Its favourite nesting places are the more open parts of the pine and birch forests where the trees are small, and separated by boggy ground into scattered groves and clumps. Beyond the limits of forest growth the Redwing breeds on the ground, generally choosing a ridge or a sloping bank for the purpose. It is not improbable that this Thrush pairs for life, as the birds are much attached to certain spots in the south and return to them yearly. Whether the same breeding localities are used each summer no naturalist appears to have taken the trouble to determine. Although the Redwing is not so gregarious in the breeding season as the Fieldfare, it is to a certain extent social, and numbers of nests may be found within a small area. The nest is usually placed in the branches of a small fir or birch, or in bushes of alder and willows, at a low altitude; very often it is built at the foot of the tree on the ground, and in treeless areas on a fence, or on the ground. It is bulky and cup-shaped. Externally it is composed of twigs, dry grass, and moss, then lined with mud, and finally lined with fine dry grass and sometimes a few roots. When disturbed at the nest the Redwing often becomes very clamorous, like its near relative the Song Thrush, fluttering to and fro in anxiety, uttering harsh grating notes.

RANGE OF EGG COLOURATION AND MEASUREMENT: The eggs of the Redwing are from four to six in number. They are bluish-green in ground colour, mottled and spotted with reddish-brown or greenish-brown. As a rule but little of the ground colour is visible, the markings covering most of the surface, and many of them are confluent. One variety has the surface colour distributed in streaky marks; another has most of the spotting confluent and in a zone round the larger end of the egg; another is clear green and almost spotless, but this latter type is exceptional. Average measurement, ·98 inch in length by ·75 inch in breadth. Incubation, performed chiefly by the female, lasts from thirteen to fifteen days.

DIAGNOSTIC CHARACTERS: The eggs of the Redwing are readily distinguished from those of all other Palæarctic Thrushes by their small s'ze, and by the markings which are normally small.

Family TURDIDÆ. Genus MERULA.
Sub-family *TURDINÆ*.

BLACK-THROATED OUZEL.

MERULA ATRIGULARIS (*Temminck*).

(British : Very rare abnormal autumn migrant.)

Number of Broods unknown. Laying season unknown.

BREEDING AREA : Eastern Palæarctic region. The Black-throated Ouzel appears to breed in the valleys of the Yenesay and the Obb from at least lat. 63° southwards, in the pine regions of Eastern Turkestan, and probably on the mountains of the Baikal area. It may also possibly breed at high elevations in the Himalayas. The breeding area of this Ouzel is very imperfectly defined.

BREEDING HABITS : Nothing is known of the breeding habits of the Black-throated Ouzel. In the late summer Mr. Seebohm met with a brood of these birds and their parents on the margin of the pine forests in Siberia, and remarked their preference for the vicinity of villages on the Yenesay—rough pastures studded with clusters of small trees. Severtzow, who states that this bird breeds in Eastern Turkestan, found it frequenting the cultivated areas, as well as the grass-clothed steppes and salt plains. The pairing habits of this Ouzel, its nest, and its habits generally during the season of reproduction have never been described.

RANGE OF EGG COLOURATION AND MEASUREMENT : The eggs of the Black-throated Ouzel have been procured on the Altai mountains, and I do not see any reason to doubt their authenticity, although it may be remarked that no scientific naturalist has yet taken them. They so closely resemble those of the Blackbird that, with the series accessible to present description, it is quite unnecessary to repeat their characteristics. The duration of the period of incubation is unknown, but it is probably chiefly performed by the female.

DIAGNOSTIC CHARACTERS : No character exists by which the eggs of this Ouzel can be separated from those of allied species, and unless thoroughly well authenticated, they are of no scientific value whatever.

Family TURDIDÆ.   Genus ERITHACUS.
Sub-family *TURDINÆ.*

ARCTIC BLUE-THROATED ROBIN.

ERITHACUS SUECICA (*Brehm*).

(British : Abnormal spring and autumn migrant.)

Single Brooded. Laying season, June.

BREEDING AREA : Northern Palæarctic region. The Arctic Blue-throated Robin, as its name implies, breeds in the regions lying north of the Arctic circle, from Scandinavia in the west to the Tchuski Land in the east, and in a similar climate at high elevations in more southern areas. Thus it is known to breed in Kamtschatka, as well as on the lofty heights of the Pamirs, and on the Himalayas.

BREEDING HABITS : The Arctic Blue-throated Robin arrives at its breeding grounds in Scandinavia early in May, but not until the end of the month in Northern Russia, and in Siberia a week or ten days later still. Its favourite breeding grounds and summer haunts are the swampy thickets of birch and willow, the clumps of underwood in pine and juniper forests, and the willow scrub and hummocky portions of the open Arctic tundra. The male is very remarkable for his varied and melodious song, often uttered whilst he hovers Pipit-like in the air, which arrests the attention at once the moment his haunts are reached. Of the pairing habits of this Robin nothing appears to be known. It may mate annually at its breeding grounds, as there is some evidence to suggest that the sexes do not migrate in company. The nest of this bird is made on the ground in some swampy spot in the forest or in one of the endless hummocks of the tundra. These mounds are a mixture of rough grass, rushes, carices, ground fruits, and dwarf willows and birches. The nest is very similar to that of the Robin, being made on the same model, placed in a hollow, the cup containing the eggs being well at the back, and the frontage to the nest being broad. Externally it is made of dry grass, roots, reindeer-moss, and dead leaves of various Arctic plants, lined with fine roots and hair. It is so cunningly concealed as to be found with the greatest difficulty, and, as is usual with species nesting in such spots, the bird sits closely. This species is not gregarious during the breeding season, but numbers of pairs may be found nesting within a small area.

RANGE OF EGG COLOURATION AND MEASUREMENT: The eggs of the Arctic Blue-throated Robin are five or six in number. They vary from bluish-green to pale olive in ground colour, somewhat indistinctly mottled with pale reddish-brown. As a rule most of the markings are pretty evenly distributed, but types occur in which the colouring matter is mostly in a cap or zone round the larger end of the egg. Average measurement, ·75 inch in length by ·55 inch in breadth. Incubation is performed chiefly if not entirely by the female, but the duration of the period is unknown.

DIAGNOSTIC CHARACTERS: The eggs of the Arctic Blue-throat cannot be distinguished from those of the Southern Blue-throat, although the two species do not inhabit the same breeding areas: locality is of first importance in identifying specimens. They also closely resemble those of the Nightingale in colour, but are much smaller.

Family TURDIDÆ. Genus SAXICOLA.
Sub-family TURDINÆ.

ISABELLINE WHEATEAR.

SAXICOLA ISABELLINA, *Ruppell.*

(British: Very rare abnormal autumn migrant.)

Double Brooded. Laying season, February, March, and June.

BREEDING AREA: Southern Palæarctic region and North-eastern Ethiopian region. The Isabelline Wheatear breeds in East Africa from Somali Land, Masai Land, and Abyssinia to Egypt, and northwards to Palestine, Asia Minor, and the steppes and plains of South-east Russia. Eastwards it breeds in Arabia, Persia, and across Siberia (south of about lat. 56°) and the plains and plateaux of Central Asia (reaching an altitude of 10,000 feet) to the upper valleys of the Amoor and North China.

BREEDING HABITS: The Isabelline Wheatear is an early migrant to its northern breeding stations, reaching those in Asia Minor towards the middle of March. Its haunts are barren grounds strewn with rock fragments and loose stones, and hillsides clothed sparingly with scrub. Mr. Danford states that in Asia Minor it frequented the fir woods. Of its pairing habits we have no information. It lives in scattered pairs during the breeding season, and evinces no gregarious instincts. Its simple song is generally uttered whilst the bird is hovering in the air. The nest of this Wheatear is always made upon the ground, and a covered site appears invariably to be chosen. It is frequently placed in the disused hole or burrow of some rodent animal, and is a cup-shaped structure, loosely put together, composed externally of dry grass and roots, and lined with fine grass, hair, and feathers. The bird, like all its kindred, is a close sitter, seldom betraying the whereabouts of the nest, which is consequently difficult to find.

RANGE OF EGG COLOURATION AND MEASUREMENT: The eggs of the Isabelline Wheatear are four or five in number. They are very pale blue in colour, and rarely exhibit any trace of pale brown markings. Average measurement, ·82 inch in length by ·65 inch in breadth. The period of incubation is unknown.

DIAGNOSTIC CHARACTERS: It is impossible to give any character by which the eggs of the Isabelline Wheatear may be distinguished from those of the Common Wheatear and other allied species. They require the most careful identification, and the locality should always be carefully noted.

Family TURDIDÆ. Genus SAXICOLA.
Sub-family TURDINÆ.

BLACK-THROATED WHEATEAR.

SAXICOLA STAPAZINA, *Vieillot.*

(British : Very rare abnormal spring migrant.)

Single Brooded probably. Laying season, April and May.

BREEDING AREA : South-western Palæarctic region. The Black-throated Chat (of which there are two fairly well-defined eastern and western races) breeds in Southern France, the Spanish Peninsula, Algeria, Morocco, Tunis, Italy, Greece, Southern Russia, Asia Minor, Palestine, and Southern Persia. I have not examined the specimen which was obtained in the British Islands, and am not prepared to say to which race it belongs. The two races appear to coalesce in Italy.

BREEDING HABITS : The Black-throated Wheatear reaches its breeding grounds in Southern Europe from the middle to the end of March. Its favourite breeding grounds are the rough uncultivated sides of mountains, sterile plains, and grass and rock-covered slopes above the vineyards and the olive groves, and below the growth of pines. This species pairs annually, soon after arriving at its summer quarters, and nest-building commences almost at once. The nest is invariably placed in a well-sheltered spot, either in a crevice of the rocks beneath the shelter of a large stone, or in a hole of a ruin. It is a cup-shaped structure, loosely and somewhat carelessly made externally of dry grass, moss, and roots, and rather more neatly lined with finer roots and sometimes hair. The bird is a close sitter, shy and wary when disturbed, and contributes little in its behaviour to the discovery of the nest. Many nests may be found within a small area, but the birds are not gregarious, and live in isolated pairs.

RANGE OF EGG COLOURATION AND MEASUREMENT: The eggs of the Black-throated Wheatear are four or five in number, usually the latter. They vary in ground colour from pale blue to dark bluish-green, marked with small spots of reddish-brown and occasionally with darker brown. Usually the markings are mostly on the large end of the egg, many of them confluent, where they form an irregular zone, but sometimes they are distributed over the entire surface. On some eggs the spots run small and very dark, on others they are blotchy and paler. Another type, less commonly seen, is almost spotless. Average measurement, ·75 inch in length by ·59 inch in breadth. Incubation is performed by the female, but the duration of the period appears not to be known.

DIAGNOSTIC CHARACTERS: The eggs of the Black-throated Wheatear require careful identification, as they can be readily confused with those of several allied species.

Family TURDIDÆ. Genus SAXICOLA.
Sub-family *TURDINÆ*.

DESERT WHEATEAR.

SAXICOLA DESERTI, *Temminck*.

(British : Rare abnormal autumn migrant.)

Number of Broods unknown. Laying season, May (?).

BREEDING AREA: South-western Palæarctic region and North-eastern Ethiopian region. The Desert Wheatear breeds in the deserts of Northern Africa from Morocco to Egypt and Nubia, northwards into Palestine, Arabia, and Southern Persia, and eastwards to the plains and plateaux of Turkestan, where it is found as high as 12,000 feet, and the mountains of Northern Cashmere.

BREEDING HABITS: The Desert Wheatear, as its name implies, is a dweller amongst the deserts, an inhabitant of sterile sandy plains, rock-strewn mountain sides, steep gorges and defiles, and the shifting, crumbling drifts of sand. How a bird can support life in many of these localities always seemed a puzzle to me. The present species appears to mate annually, is not at all gregarious, and lives in scattered pairs. Sometimes several pairs may be found dwelling in close companionship on a small area of desert ground, yet each keeps closely to itself. Notwithstanding the fact that many naturalists have met with this bird in its summer haunts, only the most meagre details have been published concerning its nidification. The nest always appears to be made upon the ground, sometimes sheltered by a little bush, sometimes in a fissure of the rocks or under a large stone, occasionally in a hole in the wall of a desert well, or even in the burrow of a marmot or other rodent. It is a loosely-made, cup-shaped structure, composed of dry grass and bents, and lined with finer grass and roots. Of the actions of the birds at the nest nothing appears to have been recorded.

RANGE OF EGG COLOURATION AND MEASUREMENT: The number of eggs laid by the Desert Wheatear is apparently unknown, the eggs themselves being very rare. Specimens that I have examined are pale greenish-blue in ground colour, with numerous small spots of rich dark brown and paler brown, usually in the form of a zone round the larger end. Average measurement, ·77 inch in length by ·49 inch in breadth. The duration of the period of incubation is unknown, as is also which parent performs the duty.

DIAGNOSTIC CHARACTERS: I am unable to give any character by which the eggs of the Desert Wheatear may be distinguished from those of the Black-throated Wheatear (possibly they are a trifle paler and the spots larger) and other allied species.

Family TURDIDÆ.   Genus MONTICOLA.
Sub-family *TURDINÆ*.

ROCK THRUSH.

MONTICOLA SAXATILIS (*Linnæus*).

(British : Very rare abnormal spring migrant.)

Said to be Double Brooded.   Laying season, April to June.

BREEDING AREA: Southern Palæarctic region. The Rock Thrush breeds in Eastern France, Southern Germany, the Spanish Peninsula, Switzerland, Austria, Turkey, Greece, Southern Russia, Asia Minor, Persia, Turkestan, and Southern Siberia as far as the Baikal area, South-east Mongolia, and Northern China. South of the Mediterranean it breeds somewhat sparingly in the mountain ranges of the Atlas.

BREEDING HABITS: The Rock Thrush is a somewhat early migrant, reaching many of its more northerly breeding places during the first half of April. The favourite haunts of this bird are mountain slopes covered with boulders and loose stones, and studded with stunted trees ; ruins, vineyards, and wild ravines up to a considerable elevation, quite up to the region of pines. Like its near allies the Redstarts, the Rock Thrush probably pairs for life ; it is known to migrate in pairs, and to arrive at its old breeding places in pairs. This species is not gregarious during the breeding season, although it travels in flocks of varying size, the companies disbanding and dispersing over a wide area as the summer haunts are reached. The nest of this species is made in a great variety of situations, but almost invariably in a covered site and well concealed. It is made under a mass of rock, or amongst heaps of loose stones, sometimes in crevices of rocks, or in holes of ruined masonry, in the walls of houses, or in trees. Exceptionally it is made under the shelter of a bush, or beneath a large tuft of drooping grass. The same variation is to be remarked in the materials. In some districts roots, moss, dry grass, and stalks of plants form the exterior, lined with hair or feathers, fine roots and grass ; in others, roots and dry grass and a few dead leaves are the only materials. In wild regions the lining of hair and feathers is rarely employed. The nest is open, cup-shaped, and rather loosely put together. The bird is a close sitter, but generally shows little anxiety when flushed from the nest unless the eggs are hatched.

RANGE OF EGG COLOURATION AND MEASUREMENT : The eggs of the Rock Thrush are four or five in number. They are a pale clear turquoise blue, sometimes faintly speckled round the larger end with pale brown, but very often entirely spotless. Frequently one egg only in a clutch will be marked, a fact which seems to show that the colour glands in this species have become almost obsolete. Average measurement, 1·0 inch in length by ·76 inch in breadth. Incubation is performed by both sexes, but the duration of the period is unknown.

DIAGNOSTIC CHARACTERS : The size and colour (especially when spotted) of the eggs, combined with the position of the nest, serve to distinguish them from those of other Palæarctic species, except from those of the Blue Rock Thrush, from which they cannot be separated. They require careful identification.

Family TURDIDÆ.   Genus ACCENTOR.
Sub-family *ACCENTORINÆ.*

ALPINE ACCENTOR.

ACCENTOR ALPINUS (*Gmelin*).

(British : Very rare abnormal autumn migrant.)

Double Brooded probably. Laying season May and July.

BREEDING AREA : South-western Palæarctic region. The Alpine Accentor breeds in the mountains of Spain, from the Sierra Nevada northwards to the Cantabrian Chain and the Pyrenees. Eastwards it breeds throughout the system of the Alps, the Apennines (including Sardinia and Sicily), the Carpathians, the mountains of Greece and Asia Minor, the Caucasus, Northern Persia, and Turkestan, in which latter country, however, the examples of this species show some affinity with *Accentor nipalensis.*

BREEDING HABITS : As soon as the higher ranges become sufficiently free from snow the Alpine Accentor quits the lower valleys and returns to its breeding places. These are the boulder-covered slopes and plateaux clothed with grass and various alpine plants and flowers, that form the zone of vegetation lying above the limits of trees and below the line of perpetual snow. It seems probable that the gregarious habits of this bird during the winter are continued more or less through the breeding season. In support of this it may be mentioned that Count Wodzicki met with colonies of Alpine Accentors on the mountains of Galizia containing as many as forty pairs. This seems incredible, and one would much like to hear further details or some confirmation of the circumstance from an independent observer. The nest of this species is invariably made upon the ground, either under the shelter of a rock or an alpine bush. It is cup-shaped, and made externally of grass stalks and roots, and lined with moss, and exceptionally with feathers, wool, or hair. Nothing appears to have been recorded of the actions of this species at the nest.

RANGE OF EGG COLOURATION AND MEASUREMENT : The eggs of the Alpine Accentor are from four to six in number. They are of a pale turquoise blue without markings of any kind. Average measurement, ·95 inch in length by ·68 inch in breadth. Incubation is probably performed by both sexes, but the duration of the period is unknown.

DIAGNOSTIC CHARACTERS: The size and colour of the eggs of the Alpine Accentor readily prevent them from being confused with the other European species. The locality is of some service too in their identification.

Family CINCLIDÆ.   Genus CINCLUS.

BLACK-BELLIED DIPPER.

CINCLUS AQUATICUS MELANOGASTER, *Brehm*.

(British : Rare abnormal autumn migrant.)

Single Brooded probably. Laying season, May and June.

BREEDING AREA: North-western Palæarctic region. The Black-bellied Dipper is not known to breed anywhere except in Scandinavia, Lapland, and Finland. The Dippers breeding in the Baltic Provinces appear to be undetermined.

BREEDING HABITS: It is not known that the habits of the Black-bellied Dipper differ from those of the common British form in any important respect. The bird is a dweller on the banks of the Scandinavian and North Russian streams, making its globular nest in crevices of the rocks, amongst the exposed roots of trees, and other suitable localities. The nest, so far as I am aware, is not known to differ from that of its southern ally, due allowance being made for locality and materials obtainable.

RANGE OF EGG COLOURATION AND MEASUREMENT: The eggs of the Black-bellied Dipper are not known to differ in number, colour, or size from those of the Common Dipper.

DIAGNOSTIC CHARACTERS: The eggs of the present form cannot be distinguished from those of the Common Dipper. The locality of the specimens should amply determine the species.

Family MUSCICAPIDÆ.   Genus MUSCICAPA.

RED-BREASTED FLYCATCHER.

MUSCICAPA PARVA, *Bechstein*.

(British : Rare abnormal autumn migrant.)

Single Brooded. Laying season, June.

BREEDING AREA: Southern Palæarctic region. The Red-breasted Flycatcher breeds in Germany, in Russia as far north as the Baltic Provinces, the Caucasus, Northern Persia, probably Turkestan (although Severtzow only records it as passing through on migration), and Southern Siberia as far as the Baikal area.

BREEDING HABITS: The Red-breasted Flycatcher

reaches the limits of its western migrations early in May. It is a shy, skulking little bird, fond of retirement, and very apt to be overlooked even in districts where it is fairly abundant, owing principally to its partiality for the summits of lofty trees. I do not find any gregarious tendency in this species during summer, each pair living isolated by themselves. The favourite nesting haunts of this Flycatcher are forests of beech and hornbeam. The bird possibly pairs for life, as many of its congeners do, and returns to one locality each year to breed. The nest is either placed in some convenient hollow in a tree-trunk or on a branch or bunch of twigs close to the stem, in just such a situation as the spotted Flycatcher so frequently selects. It is a cup-shaped structure, small and neat, and composed externally of moss studded with a few scraps of lichen, or one or two small feathers, and lined with dry grass and hair. The bird is a close sitter, often allowing itself to be taken on the nest.

RANGE OF EGG COLOURATION AND MEASUREMENT: The eggs of the Red-breasted Flycatcher are from five to seven in number. They are very pale bluish-green in ground colour, mottled and freckled with reddish-brown and with underlying markings of grayish-brown. Considerable variation occurs amongst them. On some the markings are evenly distributed over the entire surface of the egg, on others they are chiefly collected in a zone round the larger end, others are so clouded or washed with pinkish-brown as to hide almost all of the ground colour. Average measurement, ·63 inch in length by ·5 inch in breadth. Incubation is performed chiefly by the female, but the duration of the period is apparently unknown.

DIAGNOSTIC CHARACTERS: The eggs of this Flycatcher cannot readily be confused with any other European species; they somewhat closely resemble those of the Spotted Flycatcher, or even those of the Robin in colour, but are at once distinguished by their smaller size.

Family HIRUNDINIDÆ.　　　　　　　Genus PROGNE.

PURPLE MARTIN.

PROGNE PURPUREA (*Linnæus*).

(British: Very rare abnormal autumn migrant.)

Double Brooded. Laying season, April, early May, and June.

BREEDING AREA: Northern Nearctic region. The Purple Martin breeds throughout the United States of North America and in Canada, even visiting the regions above the Arctic circle. It is said also to breed in Mexico on the mountains in small numbers.

BREEDING HABITS: The spring migrations of the Purple Martin commence in the most southern localities at the end of February or early in March, the central areas are not reached until early in April, and the more northerly localities not until May. This homely American species is widely distributed throughout the northern continent in summer, and seems to frequent towns and villages as much as the quieter country districts. The nesting habits of this bird have undergone considerable change within the memory of civilized man. When houses were scarce, it used generally to breed in holes of rocks and trees, but since buildings have become common, it has deserted many of its old haunts, and taken up its residence on them. This bird probably pairs for life, and seems much attached to its mate.

The nest is a loose, slovenly structure, cup-shaped, shallow, and made externally of twigs, dry grass, straws, and leaves, and lined with feathers. Such unusual materials as rags and twine are sometimes found in the nest of this species, which is another proof of its having adapted itself to civilization. The nest is said to be thoroughly repaired and renovated before the eggs for the second brood are laid.

RANGE OF EGG COLOURATION AND MEASUREMENT: The eggs of the Purple Martin are from four to six in number. They are pure white without markings of any kind. Average measurement, ·97 inch in length by ·72 inch in breadth. Incubation is performed by the female, but the duration of the period is apparently unknown.

DIAGNOSTIC CHARACTERS: I am not aware of any character that will serve to distinguish the eggs of the Purple Martin from those of allied species.

Family CUCULIDÆ. Genus COCCYSTES.

GREAT SPOTTED CUCKOO.

COCCYSTES GLANDARIUS (*Linnæus*).

(British : Very rare abnormal spring and autumn migrant.)

Single Brooded. Laying season, April and more generally May.

BREEDING AREA: South-western Palæarctic region and North-eastern Ethiopian region. The Great Spotted Cuckoo only breeds in Europe in the Spanish Peninsula. Elsewhere it breeds in Northern Africa from Morocco to Egypt and Nubia, and northwards in Palestine, Asia Minor, and Northern Persia.

BREEDING HABITS: The migration of the Great Spotted Cuckoo into Europe begins early in March, and continues through that month into April, and about the same dates apply to Palestine, although in Asia Minor the bird is said not to arrive until the end of March. The migration lasts through April in some localities. This Cuckoo is gregarious on passage, journeying in flocks. The haunts of the Great Spotted Cuckoo are well-timbered localities, wooded districts, especially where the timber occurs in groves. The bird is no nest-builder, and, like the Common Cuckoo, deposits its eggs in the nests of other species, to whom all care of the young is transferred. Although this habit had long been suspected, to Brehm must be given the credit of its absolute discovery, which he made in Egypt in 1850. Some species of Crow is usually selected by this Cuckoo to play the part of foster-parent. The selected species varies considerably according to locality. Thus in Egypt the Hooded Crow is chosen, in Palestine the Black-headed Jay, in Algeria the Moorish Magpie and perhaps the Little Owl, and in Spain usually the Common Magpie. Nothing appears to be known respecting the pairing habits of this bird. The hen Cuckoo carries her egg in her beak, and inserts it in the selected nest, which seems a very unnecessary proceeding, seeing that the nests used by this species are easily accessible, and seems to be an inherited habit probably common to all parasitic Cuckoos. Perhaps the habit has arisen through motives of safety, the bird running much less chance of detection from the rightful owners of the nest during the few moments taken up in dropping an egg from her bill, than she would otherwise incur if she sat for some time on the nest to lay it there in the usual way.

RANGE OF EGG COLOURATION AND MEASUREMENT:

It is difficult to say the number of eggs usually produced by each individual Cuckoo every season, but sometimes as many as four are laid in the same nest (whether by one female or several is not known), but usually only two, and in some cases but one. They are pale bluish-green in ground colour, spotted and blotched with pale brown, and with underlying markings of lilac-gray. They are subject to some local variation, but to nothing near the extent presented in the eggs of the Common Cuckoo. On some varieties the surface spots are small and few, but the underlying markings are numerous and conspicuous; on others most of the markings are collected in a zone round the larger end of the egg, and take the form of streaks and scratches with a few spots between. The markings on most eggs are pretty generally distributed over the surface, but are slightly more numerous round the larger end. Average measurement, 1·2 inch in length by ·92 inch in breadth. Incubation is performed by the foster-parent, but the duration of the period is unknown.

DIAGNOSTIC CHARACTERS: The eggs of this Cuckoo very closely resemble certain types of those of the Common Magpie, but are readily distinguished by their smoother grain, greater rotundity, smaller size, and reddish-brown (not olive-brown) surface spots.

Family CUCULIDÆ. Genus Coccyzus.

BLACK-BILLED CUCKOO.

Coccyzus erythrophthalmus (*Wilson*).

(British : Very rare abnormal autumn migrant.)

Single Brooded. Laying season, April, May, and June.

BREEDING AREA: Eastern Nearctic region. The Black-billed Cuckoo breeds in suitable districts throughout the North American Continent east of the Rocky Mountains, as far north as Labrador, and as far south as Georgia and Texas.

BREEDING HABITS: The Black-billed Cuckoo reaches its more northerly breeding places in May, but its eggs appear to be laid by that date in the southern districts. It is a shy, seclusion-loving bird, and its favourite haunts are woods and thickets, and the belts of timber on the banks of streams. This Cuckoo is not parasitic, but builds a nest and hatches its eggs in the normal way; whether it pairs annually or for life appears not to be known. It is not gregarious during the breeding season, but lives in scattered pairs. The nest is usually placed on the flat, horizontal branch of a tree at some height from the ground, or in the centre of a dense bush—a thorn-bearing one being preferred. It is a well-made structure, resembling the coarse architecture of the crows, rather flat and shallow, yet open, and composed externally of coarse and fine twigs and roots, and lined with finer roots and grass. The bird makes little or no demonstration when flushed from the eggs, and soon conceals itself in the surrounding cover.

RANGE OF EGG COLOURATION AND MEASUREMENT: The eggs of the Black-billed Cuckoo are from three to five in number. They are somewhat rough in grain, without gloss, and uniform pea-green in colour without markings. Average measurement, 1·1 inch in length by ·85 inch in breadth. The duration of the period of incubation is unknown, also which parent performs the task.

DIAGNOSTIC CHARACTERS: The eggs of the present species may generally be distinguished from those of the Yellow-billed Cuckoo, by their much smaller size and darker colour; but as small eggs of the latter species cannot be separated, the greatest care is required in their identification.

Family CUCULIDÆ. Genus COCCYZUS.

YELLOW-BILLED CUCKOO.

COCCYZUS AMERICANUS (*Linnæus*).

(British : Very rare abnormal autumn migrant.)

Single Brooded. Laying season, April, May, and June.

BREEDING AREA: Southern Nearctic region. The Yellow-billed Cuckoo breeds throughout the continent of North America in suitable districts from New Brunswick and Minnesota. in the north, southwards to Mexico and the West Indies. It is, however, less common in the northern areas.

BREEDING HABITS: The Yellow-billed Cuckoo arrives at the northern limits of its migrations towards the end of April, but is a month or more earlier in the south. This bird is also shy and secretive in its habits, keeping much amongst the dense foliage, and is far more often heard than seen. Its favourite haunts are woods and clumps or "bluffs" of trees, but the bird may also be seen in orchards and well-wooded parks. Although it migrates in flocks to and from its breeding grounds, it is not in any way gregarious during the nesting season, but lives in scattered pairs. As the males are said to arrive before the females, this Cuckoo probably pairs annually. It is not parasitic in its habits, but provides for and rears its young in the normal way. The nest is either placed in a tree or in a large dense bush, and is carelessly and loosely made, flat and shallow but open, composed externally of twigs and roots, and lined with dry grass and finer roots. The male is said to keep close watch in the neighbourhood of the nest, and to be very jealous and pugnacious in beating off intruding birds or animals.

RANGE OF EGG COLOURATION AND MEASUREMENT: The eggs of the Yellow-billed Cuckoo are from three to five in number. They are without gloss, rough in texture, almost like the eggs of a Heron, and of a clear pea-green colour without markings. Average measurement, 1·25 inch in length by ·92 inch in breadth. Incubation appears to be performed chiefly if not entirely by the female, but the duration of the period is unknown.

DIAGNOSTIC CHARACTERS: As the eggs of this Cuckoo and those of the preceding species overlap in size, no character can be given which will distinguish them in every case, so that they require careful identification. The eggs of the Yellow-billed Cuckoo are generally much the largest in size and palest in colour.

Family CYPSELIDÆ. Genus CYPSELUS.
WHITE-BELLIED SWIFT.
CYPSELUS MELBA (*Linnæus*).
(British : Rare abnormal spring and autumn migrant.)
Single Brooded. Laying season, May.

BREEDING AREA : Southern Palæarctic, North-eastern Ethiopian, and Oriental regions. The White-bellied Swift breeds on the mountains of Eastern France, the Pyrenees, and mountains of Spain, the Alps, the Carpathians, Apennines, and other high ranges of South and Central Europe, to the Urals, where it goes as far north as lat. 55°. Eastwards it breeds in the mountains of Asia Minor, Palestine, Persia, Turkestan, the Himalayas, the mountains of Western India, and Ceylon. South of the Mediterranean it breeds throughout the Atlas range and in the mountains of Abyssinia.

BREEDING HABITS : The White-bellied Swift begins to cross the Mediterranean during the latter half of March, and does not reach its more northerly breeding places before May. This bird is more or less gregarious during the breeding season, nesting in scattered colonies ; it is also social, and in many districts fraternizes with the Common Swift, the two species rearing their young in the same colony. Its favourite breeding haunts are in the mountains, amongst lofty cliffs, steep gorges, and the precipitous sides of ravines and passes. It frequents buildings in mountainous areas, however, to some extent, one of its most famous nesting places being in the cathedral at Berne, and others occur on lofty buildings in the Alps. I met with colonies of this fine Swift on the towering precipices at Constantine, also in the gorge at El Kantara, most of its nests being quite inaccessible to man. The White-bellied Swift pairs for life, and yearly returns to its old nesting places. The nest is always made in a covered site, either in a cranny or a cleft of the cliffs, or in the tower of a cathedral or a mosque, and when in the cliffs is generally inaccessible, being as far up the cleft as possible. The nest is flat and saucer-shaped, and is composed of dry grass, scraps of paper and fir bark, bits of straw, and dead leaves, all more or less cemented together and covered with the dry saliva or mucous fluid which must of necessity get upon them whilst being conveyed to the hole and arranged in place; the lining is of feathers. Nests are often placed close together. Hume describes a cluster of three or four nests grouped together in one solid mass, in a series of chambers; whilst Mr. Wilson, who recently examined the colony of this species in Berne Cathedral, found several nests within a space of three yards on the same beam. The Rev. H. A. Macpherson, who has also visited this colony, states that he noticed green grass in many of the nests. Canon Tristram has found a nest in a cave on Mount Gilead which appeared to have been made in the deserted nest of a Rock Nuthatch (*Sitta syriaca*).

RANGE OF EGG COLOURATION AND MEASUREMENT: The eggs of the White-bellied Swift are from two to four in number, but the usual clutch is two, the larger number probably being made up by two females. They are very long and narrow, without gloss, almost chalky in texture, and pure white. Average measurement, 1·2 inch in length by ·77 inch in breadth. Incubation is performed by both sexes, the male relieving the female at the nest, but the duration of the period is apparently unknown.

DIAGNOSTIC CHARACTERS: The large size of the eggs of this Swift prevents them from being confused with those of any other European species.

Family CAPRIMULGIDÆ.   Genus CAPRIMULGUS.
RED-NECKED NIGHTJAR.
CAPRIMULGUS RUFICOLLIS, *Temminck*.
(British : Very rare abnormal autumn (?) migrant.)
Single Brooded probably.   Laying season, May and June.

BREEDING AREA : South-western Palæarctic region. The Red-necked Nightjar breeds in the southern half of the Spanish Peninsula, and in North-west Africa from Morocco and the Canary Islands to Tunis.

BREEDING HABITS : But little has been recorded of the habits of the Red-necked Nightjar. It is a regular summer visitor to Spain, and frequents wooded and scrub-covered districts, but so far as can be determined, there is nothing strikingly different in its economy from that of the common British species. Of its pairing habits nothing appears to be known. The nest—if such it can be called—is merely a hollow in the ground.

RANGE OF EGG COLOURATION AND MEASUREMENT : The eggs of the Red-necked Nightjar are two in number, and with the exception of being on an average slightly less spotted, do not differ from those of the Common Nightjar in colour. Average measurement, 1·3 inch in length by ·87 inch in breadth. Incubation period unknown ; incubating sex also unknown.

DIAGNOSTIC CHARACTERS : The eggs of this species may usually be distinguished from those of the Common Nightjar by their slightly larger size and less amount of marking. No character is known by which they may be completely diagnosed.

Family CAPRIMULGIDÆ.   Genus CAPRIMULGUS.

EGYPTIAN NIGHTJAR.

CAPRIMULGUS ÆGYPTIUS, *Lichtenstein.*

(British : Very rare abnormal spring migrant.)

Single Brooded probably.   Laying season, June and July.

BREEDING AREA: South-central Palæarctic and North-eastern Ethiopian regions. The Egyptian Nightjar breeds in Nubia, Egypt, probably Arabia, Baluchistan, and Western Turkestan.

BREEDING HABITS : The Egyptian Nightjar is another species whose habits have been little studied by naturalists. It is a bird of the desert regions, and, due allowance being made for the difference of habitat, closely resembles the Common Nightjar in its economy and movements. Its favourite haunts are sandy plains covered with scrubby vegetation. On migration it is more or less gregarious, but during the breeding season lives in scattered pairs. It appears to pair annually, and the males are the first to arrive at the nesting grounds. The nest is a mere depression in the sand, either amongst esparto grass, or under the shelter of an acacia or tamarisk bush. The bird is a close sitter, as might naturally be expected in a species so closely resembling in colour the ground on which it rests, but when flushed has been observed to run from the shelter of one bush to that of another, with throat extended and uttering a croaking note.

RANGE OF EGG COLOURATION AND MEASUREMENT: The eggs of the Egyptian Nightjar are two in number, and resemble closely those of the Common Nightjar, but according to Von Heuglin, who met with this bird breeding on some sandy islands in Dongola, they are a little smaller and paler and of a more yellowish hue. The latter character is probably a protective one, and renders the eggs in close harmony with the tints of the sand on which they rest. Average measurements unknown.

DIAGNOSTIC CHARACTERS: As our information respecting the eggs of the Egyptian Nightjar is so meagre, it is quite impossible to say whether any of the characters they present are diagnostic. Information is much to be desired.

Family MEROPIDÆ. Genus MEROPS.

BEE-EATER.

MEROPS APIASTER, *Linnæus*.

(British : Rare abnormal spring and autumn migrant.)

Single Brooded. Laying season, end of April to first half of June.

BREEDING AREA: Southern Palæarctic region and North-eastern Ethiopian region. The Bee-Eater only breeds exceptionally in Europe north of lat. 50° in the west and lat. 52½° in the east. It breeds commonly in the Spanish Peninsula, much less so in the extreme south of France, but is abundantly distributed throughout Southern Europe below the limits already given in Italy, Austria, the valley of the Danube, Turkey, Greece, and Southern Russia. It also breeds in Asia Minor, Palestine, Persia, Turkestan, and, it is said, in the Altai Mountains. South of the Mediterranean it breeds in Morocco, Algeria, Tunis, and Egypt. It is also said to breed in the Canaries and Madeira ; and according to Layard in the Cape Colony.

BREEDING HABITS: The Bee-Eater crosses the Mediterranean to its more northerly breeding haunts during the whole of April and the early part of May. It is possible that this beautiful bird pairs for life; I remarked it in pairs on migration, and the same nesting places are frequented year by year. It is also a very gregarious bird, and breeds in colonies of varying size. The favourite resorts of this Bee-Eater are river valleys and the vicinity of earth cliffs in open country; the bird also frequents deserted fortifications and earthworks. This bird lays its eggs in burrows or holes, which it excavates in some suitable bank just like a Sand-Martin or a Kingfisher, and a new hole is apparently made each season. Many burrows are made close together in the face of a single cliff, the colony being contracted or scattered according to the extent of available bank. The burrow is made by both birds, each working in turn with bill and feet, and sometimes extends as much as nine feet into the solid bank! As a rule the burrows are from two to four feet in depth, straight and almost horizontal, but occasionally they are very tortuous, and one communicates with another by a narrow gallery. In Spain the burrows, in some cases, where cliffs are wanting, are sunk into the level ground almost perpendicular or in an oblique direction. At the end of the tunnel a little chamber is formed, but no nest is provided, and the eggs are deposited on the bare ground or on the wing cases of insects, the refuse of the bird's food. The Bee-Eater is a close sitter, and usually allows itself to be removed from its eggs. The scene outside a colony is a very pretty one, the birds entering and leaving their holes and skimming about, displaying their rich plumage to best advantage in the brilliant sun.

RANGE OF EGG COLOURATION AND MEASUREMENT: The eggs of the Bee-Eater are from five to eight or even nine in number. They are very rotund, highly polished, and pure white. Average measurement, 1·0 inch in length by ·9 inch in breadth. Incubation is performed chiefly by the female, but the duration of the period appears to be unknown.

DIAGNOSTIC CHARACTERS : I know of no character which will distinguish the eggs of the Bee-Eater from those of certain allied species. They require careful identification.

Family CORACIIDÆ. Genus CORACIAS.

ROLLER.

CORACIAS GARRULUS, *Linnæus.*

(British : Rare abnormal spring and autumn migrant.)

Single Brooded. Laying season, May and first half of June.

BREEDING AREA: South-western Palæarctic region and North-western Oriental region. The Roller breeds throughout Europe, south of about lat. 60°, with the exception of Denmark (where it is rare), the Netherlands, Belgium, and Northern France. It becomes most abundant in the countries bordering the Mediterranean, Black, and Caspian Seas. Eastwards it breeds in Asia Minor, Palestine, Persia, Turkestan, Afghanistan, Cashmere, North-west India (the Punjaub), and Southern Siberia, as far north as Omsk and as far east as the Altai Mountains. South of the Mediterranean it breeds in North-west Africa, from Morocco to Tunis.

BREEDING HABITS : The migration of the Roller across the Mediterranean northwards into Europe takes place at about the same time as that of the Bee-eater, and lasts through April to the beginning of May. On migration the Roller is gregarious, and continues gregarious for some time after arriving at its breeding grounds. These flocks eventually disband into pairs, and scatter themselves over the country, but Canon Tristram asserts that he met with this bird breeding in a colony in Palestine. It is probable that the Roller pairs for life, but whether it returns to one particular spot to breed each season I am unable to say. The favourite breeding grounds of the Roller are open woodlands, and broad plains studded with clusters of trees. I found it in Algeria most partial to park-like country; it is said also to frequent river valleys, where the banks are steep. During the love season the Roller often indulges in curious aërial evolutions, something after the manner of a Tumbler Pigeon, male and female chasing each other, and very noisy. The Roller makes its nest in a hole of a tree, or in the crevice of a rock, but holes in banks or walls are almost as frequently selected. When in trees, no nest is made, and the eggs rest upon the powdered wood at the bottom, but when in other situations a slight nest of dry grass, twigs, roots, and occasionally a few feathers is formed. The bird does not always excavate the hole itself, for in Algeria the timber was full of suitable hollows ready made, but in many cases it certainly does so. The Roller is a close sitter, often allowing itself to be dragged out of its retreat without any attempt to escape.

RANGE OF EGG COLOURATION AND MEASUREMENT: The eggs of the Roller are from four to six in number, sometimes only three. They are globular in form, highly polished, and pure white. They vary a good deal in size and shape. Average measurement, 1·4 inch in length by 1·1 inch in breadth. Incubation, performed by both sexes, lasts from eighteen to twenty days.

DIAGNOSTIC CHARACTERS: It is impossible to give

any character by which the eggs of the Roller may be distinguished from those of several allied species. Careful identification is required.

Family ALCEDINIDÆ.   Genus CERYLE.

BELTED KINGFISHER.

CERYLE ALCYON (*Linnæus*).

(British: Very rare abnormal autumn migrant.)

Partially Double Brooded. Laying season, April, May, and June.

BREEDING AREA: Nearctic region. The Belted Kingfisher breeds throughout the North American continent from the Atlantic to the Pacific, north to the Arctic Ocean, and south to Central America.

BREEDING HABITS: The favourite haunts of the Belted Kingfisher are streams, lakes, and ponds whose banks are sufficiently wooded or clothed with vegetation to afford cover and concealment, and steep enough to furnish a requisite breeding place. It is unsociable and lives a solitary life, each pair having particular beats to which they closely keep. There can be little doubt that this Kingfisher pairs for life, and resorts to a chosen nesting place year after year. Like the Common Kingfisher of our English streams, it rears its young in a hole in a bank. This hole is excavated by the birds themselves, each working in turn, and often so rapidly that it is completed in a single night. The burrow is usually from two to four feet in depth, but instances have been known where it has extended as much as fifteen feet into the bank. As a rule it is pretty straight, slightly curving upwards, but is sometimes more tortuous, according to the state of the ground which has to be pierced. At the end of the burrow a small chamber is formed. It has been said that this species makes a slight nest of dry grass, but the probability is the eggs are deposited on a layer of fish bones, scales, etc., the remains of the bird's food. Nests are not always made near water, and the bird has been known to breed upwards of a mile from it. The bird is much attached to its nest-hole, is a close sitter, and has even been known to feign being wounded to allure an intruder from its helpless young.

RANGE OF EGG COLOURATION AND MEASUREMENT: The eggs of the Belted Kingfisher are six or seven in number. They are very rotund in shape, glossy in texture, and pure white. Average measurement 1·35 inch in length by 1·05 inch in breadth. Incubation, performed by both sexes, lasts, on the authority of Audubon, sixteen days, but I should feel disposed to put it more probably at four or five days more.

DIAGNOSTIC CHARACTERS: There is no character by which the eggs of this Kingfisher can be distinguished from those of the Roller, the Pied Kingfisher, or the Smyrna Kingfisher, but of course these species do not breed in America. Locality is therefore of some service in identifying them.

Family STRIGIDÆ. Genus NYCTALA.
Sub-family BUBONINÆ.

TENGMALM'S OWL.

NYCTALA TENGMALMI (*Gmelin*).

(British : Rare nomadic autumn migrant.)

Single Brooded. Laying season, May and June.

BREEDING AREA: Northern Nearctic and Palæarctic regions. Tengmalm's Owl breeds in the pine regions south of the Arctic circle in Europe and Asia. It breeds in Northern Scandinavia, in Lapland, Finland, and Russia, to at least as far south as Orenberg. It also breeds in the Alps and the Carpathians, but is not yet proved to do so in the Pyrenees. Herr Hartert states that it may probably breed in North-eastern Prussia. Eastwards it breeds in the pine forests of Siberia to the Pacific. In the New World it breeds right across the continent, as far north as the Arctic circle, from Alaska to Labrador.

BREEDING HABITS: Tengmalm's Owl is for the most part a resident in the pine forests of the Arctic regions. But little is known of its nesting economy, and for what little information we do possess we are indebted to the researches of Wolley and Wheelwright, who met with this bird in Lapland. Of its pairing habits nothing definite is known, but it probably mates for life as so many other Owls are known to do. This Owl appears not to make any nest, but to take possession of the deserted hole of a Black Woodpecker or other cavity in a tree trunk : whilst Wolley found it occupying the hollowed logs which the peasants place for the Golden-eyes to breed in. No nest of any kind is made, and the eggs rest upon the powdered wood or the layer of pellets and food refuse cast up by the parent birds.

RANGE OF EGG COLOURATION AND MEASUREMENT: The eggs of Tengmalm's Owl are from four to seven in number. Instances are on record where as many as ten have been found, but these were probably the produce of two females. They differ a good deal in shape, some being much more elongated than others, and are pure white and somewhat smooth in texture. Average measurement, 1·28 inch in length by 1·1 inch in breadth. The duration of the period of incubation is unknown, as is also the sex which undertakes it.

DIAGNOSTIC CHARACTERS: The eggs of Tengmalm's Owl, if the *locality* can be relied upon, are not easily confused with those of any other species, but unless they are carefully identified and authenticated they are worthless.

Family STRIGIDÆ.　　　　　　　　　　Genus ATHENE.
Sub-family *BUBONINÆ*.

LITTLE OWL.

ATHENE NOCTUA (*Scopoli*).

(British : Rare abnormal migrant.)

Single Brooded.　Laying season, latter half of April and first half of May.

BREEDING AREA: South-western Palæarctic region. The Little Owl breeds throughout Europe south of Scandinavia and lat. 56° in Russia, becoming most abundant in the south, especially in those countries bordering the Mediterranean.

BREEDING HABITS: The haunts of the Little Owl are as much in the wooded cultivated districts near villages and farm-houses, as in the more desolate portions of the mountain sides and uninhabited plains. It frequents olive groves and vineyards as well as most of the classic ruins in Italy and Greece. There is no gregarious tendency in this species, and it lives in scattered pairs. It probably pairs for life, and continues in many cases to haunt one particular spot in which to roost and to breed year after year. As is usual, the nesting place is generally the daily retreat as well. The eggs are laid in hollow trees, in crevices of rocks, in holes of buildings and ruins, and exceptionally under the exposed roots of a tree. In Algeria I have found nests of the southern race of this Owl (*Athene glaux*) under a rock boulder on the mountain side, and in a hole in a low range of mud cliffs. Little or no nest is ever made, the eggs resting upon any dust or refuse that may chance to be in the hole, or upon a layer of pellets containing food refuse.

RANGE OF EGG COLOURATION AND MEASUREMENT: The eggs of the Little Owl are from four to six in number, rarely only three. They are oval in shape, somewhat smooth in texture, and white. Average measurement, 1·4 inch in length by 1·15 inch in breadth. Incubation, performed chiefly by the female, is said, on the authority of Mr. Meade-Waldo, to last twenty-eight days.

DIAGNOSTIC CHARACTERS: There is no thoroughly reliable character by which the eggs of the Little Owl may be distinguished from those of the Scops Owl (the only European species with which they can be confused); they are slightly larger on an average, not quite so rotund in form, and the texture is a little coarser.

Family STRIGIDÆ.   Genus NYCTEA.
Sub-family BUBONINÆ.

SNOWY OWL.

NYCTEA NYCTEA (*Linnæus*).

(British : Fairly regular nomadic autumn migrant.)

Single Brooded. Laying season, end of May or in June.

BREEDING AREA : Northern Nearctic and Palæarctic regions. The Snowy Owl is principally confined during the breeding season to the country north of the Arctic circle. It breeds on the fells of Norway and Sweden, in Lapland, North Russia, and Nova Zembla. Eastwards in Asia it breeds on the tundras in the most northerly parts of Siberia to the Pacific. In the New World it is only known to breed in the extreme north from Alaska to Labrador, and northwards to Grinnell Land, where it was observed nesting by Col. Feilden in lat. $82\frac{1}{2}°$.

BREEDING HABITS : The Snowy Owl wanders little from its home during winter, retiring perhaps as far south as the forests, but returning with the advent of summer. Its only breeding haunts are the fells and tundras and barren grounds that stretch from the limits of forest growth northwards to the coasts and islands of the Arctic Sea. It is not in the usual sense a gregarious bird, living for the most part solitary, except during the short period of the breeding season ; but flocks of this Owl have from time to time been met with far at sea. Of the pairing habits of this Owl nothing is recorded. It probably mates for life, but the two birds only live in close companionship during reproduction. The nest of the Snowy Owl is either placed on the ground, or on a ledge of a cliff, either overlooking a river or the sea, or on some slight mound on the tundra. It is merely a hollow trampled down in the soil or moss, in which a few feathers and pellets accumulate and form the only bed on which the eggs repose. The eggs in most cases appear to be laid in pairs at intervals, the bird commencing incubation at once, so that by the time all are deposited, some of the earlier ones may be hatched, and the young assist in incubating the rest. Eggs and young birds in various stages of growth may therefore be found at the same time in one nest. The male bird keeps jealous watch near the nest ready to defend it and beat off any intruding creature, or to warn the female of the approach of danger. She sits lightly, and is up at the least alarm, the two birds careering wildly about round the spot, screaming incessantly.

RANGE OF EGG COLOURATION AND MEASUREMENT: The eggs of the Snowy Owl are usually from six to eight in number, but ten have been found. They are rather rough in grain, show but little polish, and are white, often with a scarcely perceptible yellow tinge. Average measurement, 2·3 inches in length by 1·7 inch in breadth. Incubation, performed by the female, lasts (in confinement) thirty-two days.

DIAGNOSTIC CHARACTERS: There is absolutely no reliable character by which the eggs of the Snowy Owl may be distinguished from those of the Eagle Owl (the only species with which they can be confused in Europe). Generally they are a little smaller than that bird's, which also never show any yellow tint, and are more oval in shape. The breeding grounds of the two species are also quite different.

Family STRIGIDÆ. Genus SURNIA.
Sub-family *BUBONINÆ.*

HAWK OWL.

SURNIA FUNEREA (*Linnæus*).

(British : Very rare nomadic winter migrant.)

Single Brooded. Laying season, latter half of April to end of June.

BREEDING AREA : Northern Nearctic and Palæarctic regions. The Hawk Owl is divisible into three fairly well-defined races, inhabiting Europe, Siberia, and America respectively, but for the purposes of the present article I have deemed it most expedient to treat them as one. It breeds in the pine forests of Scandinavia and North Russia, occasionally inhabiting the birch woods on the borders of the tundras. It is said to breed as far south in Russia as the Governments of Moscow and Smolensk, and in the Urals as low as Orenburg. East of the Urals it breeds in similar localities across Siberia, south to Northern Turkestan and the Amoor. In the New World it breeds in the Arctic regions from Alaska to Labrador and Newfoundland.

BREEDING HABITS : The Hawk Owl is practically a resident in its northern forests, only undertaking such nomadic wanderings as scarcity of food may cause. Its favourite breeding grounds are the Arctic pine forests, but it is also partial to the forests of birch, especially in localities where the timber is old and large. It is not a gregarious species at any season, and for the most part lives a solitary life except during the breeding season, when the probably life-mated pair live in closer company. This Owl, at least in the Palæarctic region, is not known to make any nest but to deposit its eggs in a hole in a tree, especially in the deserted nest of a Black Woodpecker, or in the hollowed logs and the nest-boxes placed for the accommodation of Ducks by the peasants. It has also been known to lay its eggs on the top of a broken pine trunk, in which a female Golden-eye was sitting on her nest some six feet below; the top of a broken birch stem has also been selected. In America, the Hawk Owl is stated by Macfarlane to build a nest of sticks, and the observation is confirmed by later observers. Mr. Raine, in his interesting work on *Bird-nesting in North-west Canada*, gives the date and locality of at least three nests, two of which were made of "sticks and leaves," and one of them lined with grass and feathers. Further information is much to be desired, as it seems probable that the nests were those of some other birds which the Owls had appropriated. These birds are very pugnacious when disturbed at the nest. The eggs appear often to be laid at intervals, and to be incubated at once.

RANGE OF EGG COLOURATION AND MEASUREMENT: The eggs of the Hawk Owl are from five to eight in number, smooth in grain, with some polish, and pure white. Average measurement 1·55 inch in length by 1·2 inch in breadth. Incubation is performed by both sexes, but the duration of the period is apparently unknown.

DIAGNOSTIC CHARACTERS: The eggs of the Hawk Owl unfortunately cannot be distinguished from those of the Long-eared Owl and the Short-eared Owl, so that they require the most careful identification to render them of any scientific value.

Family STRIGIDÆ. Genus SCOPS.
Sub-family BUBONINÆ.

SCOPS OWL.

SCOPS SCOPS (*Linnæus*).

(British : Rare abnormal spring and autumn migrant.)

Single Brooded. Laying season, May.

BREEDING AREA : South-western Palæarctic region. The Scops Owl breeds in Central and Southern Europe, but becomes most abundant in the countries bordering the Mediterranean. It breeds in Switzerland, Southern France, the Spanish Peninsula, Italy, Austria, Turkey, the Danubian provinces, Greece, and Southern Russia. It also breeds in Asia Minor, Palestine, Persia, and Turkestan. South of the Mediterranean it breeds in West and North-west Africa, but these examples are said to be smaller and possibly sub-specifically distinct.

BREEDING HABITS : This pretty little Owl is for the most part a migrant north of the Mediterranean, arriving from its African winter haunts about the end of March or early in April. As it returns season by season to its old nesting place, there can be no doubt that the Scops Owl pairs for life. It is not at all gregarious, and lives in scattered pairs, but it may be very common in certain districts where food and shelter are abundant. The favourite haunts of this Owl are olive groves, vineyards, gardens, and groves of trees even in large towns. It also frequents the wilder and less cultivated districts, ascending the mountains to at least as far as the pine zone. In Africa the cork woods are a favourite retreat. This Owl also makes no nest, but lays its eggs generally in a hole in a tree, less frequently in a hole in a wall, on the dust or refuse of its food that may by chance have accumulated there. The bird is a close sitter, like all hole-building species, and generally allows itself to be taken from the nest.

RANGE OF EGG COLOURATION AND MEASUREMENT: The eggs of the Scops Owl are five or six in number. They are somewhat rotund in shape, smooth in texture, and pure white. Average measurement, 1·25 inch in length by 1·05 inch in breadth. Incubation is performed by the female, but the duration of the period is apparently unknown.

DIAGNOSTIC CHARACTERS: There is no reliable character by which the eggs of the Scops Owl may be distinguished from those of the Little Owl (the only European species with which they can be confused): they are a little smaller on an average, more globular, and the grain is perhaps a trifle finer.

Family STRIGIDÆ.   Genus BUBO.
Sub-family *BUBONINÆ*.

EAGLE OWL.

BUBO MAXIMUS, *Gerini*.

(British : Rare nomadic winter migrant.)

Single Brooded.   Laying season, March and April.

BREEDING AREA: Palæarctic region. The Eagle Owl (typical form) breeds throughout Europe in suitable localities from Scandinavia and Northern Russia southwards to the Mediterranean, and beyond that sea in Africa north of the Atlas. East of the Ural Mountains in Siberia, Persia, Afghanistan, the Himalayas, and Turkestan, a pale race of Eagle Owl is found (*B. sibiricus*);

but beyond this area the typical form re-appears in the Amoor, China, and Japan.

BREEDING HABITS: The migrations of the Eagle Owl are extremely limited, and are merely nomadic wanderings in quest of food. It is not gregarious nor social, and lives principally alone except during the breeding season. This Owl probably pairs for life, and apparently continues to haunt a certain spot for years, the daily retreat being as usual the breeding place too. The haunts of the Eagle Owl are large forests, especially such where the trees are for the most part non-deciduous. In wild, uncultivated localities it shows much preference for mountain forests and woods, in which rocks containing caves and hollows occur. This Owl also never makes a nest for itself, but either takes possession of a deserted nest of an eagle or other large bird, or selects a convenient ledge on a cliff. Less frequently it has been known to breed upon the ground at the foot of a tree, probably because a convenient nest was not to be found; and Wolley had the eggs from under the roots of a fallen tree in Lapland. The eggs are laid upon a slight heap of pellets and food refuse, the bird apparently doing nothing in the way of providing accommodation for them, beyond scraping or treading a slight hollow, when an old nest is not used.

RANGE OF EGG COLOURATION AND MEASUREMENT: The eggs of the Eagle Owl are usually two in number and never exceed three. They are rotund in shape, rather coarse in grain, and pure white. Average measurement, 2·3 inches in length by 1·9 inch in breadth. Incubation lasts from thirty-four to thirty-six days in confinement, where this species frequently breeds, and is doubtless of the same duration when the bird is at liberty. The female probably performs most if not all of the task, but on this authorities are silent.

DIAGNOSTIC CHARACTERS: The eggs of the Eagle Owl can only be confused with those of the Snowy Owl in Europe. They are on an average slightly larger and more globular, never also displaying any yellow tinge. The breeding grounds of the two species do not impinge.

Family VULTURIDÆ. Genus GYPS.

GRIFFON VULTURE.

GYPS FULVUS (*Brisson*).

(British : Very rare abnormal migrant.)

Single Brooded. Laying season, January to March.

BREEDING AREA: South-western Palæarctic region and North-eastern Ethiopian region. The Griffon Vulture breeds in the mountainous districts of the Spanish Peninsula, in the Pyrenees, the Alps, the mountains of Sardinia and Sicily, the Carpathians, the mountains of Turkey, Greece, and Asia Minor, and in the Caucasus and Southern Urals. Eastwards it breeds in Persia and Turkestan; whilst south of the Mediterranean it is a common bird in Africa north of the Great Desert from Morocco to Egypt, and southwards into Nubia.

BREEDING HABITS: Although the Griffon Vulture wanders far and wide over the countries it inhabits, during the nesting season it is practically confined to the neighbourhood of precipices, the steep rugged slopes of mountains and ranges of cliffs, especially those of a limestone formation, and full of hollows and caves. There can be no doubt that the Griffon Vulture pairs for life, and continues to breed in one particular place for years in succession. This Vulture is gregarious and breeds in colonies, the nests being scattered here and there amongst the cliffs which the birds frequent. The nest, which is patched up and often added to each year, is often a massive bulky structure, and is generally made on the floor of a little cave near the entrance, or at the bottom of some wide hollow, especially in such spots where the cliffs overhang and are covered above and below with a dense and impenetrable growth of aloes, prickly pears, and other vegetation. It is made of sticks, twigs, and branches of trees, and lined with dry grass, leaves, and dead palmettoes. Some nests are very neatly finished, the bowl or cup containing the eggs being some fifteen inches across and four or five inches deep. The newer nests are usually not so elaborate nor so neatly finished as the older ones, which are the work of years. All round about the place where the nest is situated is splashed with the white droppings of the birds, and a sickly often almost unbearable stench pervades the whole colony. When disturbed the birds rise into the air and wheel about, occasionally swooping past their nests, but never showing the slightest inclination to attack the intruder. If the first eggs are taken others are generally laid.

RANGE OF EGG COLOURATION AND MEASUREMENT: The eggs of the Griffon Vulture are rarely two in number and generally only one. They are coarse in grain, have little or no polish, and are generally white without markings, save perhaps a few nest- or blood-stains. Many eggs, however, are somewhat sparingly marked with genuine colour pigment. I have examined many of these spotted eggs in the National and other collections. Some eggs are sparingly streaked with grayish-brown, others are blotched and spotted with reddish-brown over most of the surface, whilst others yet again are irregularly zoned round the end with deep reddish-brown spots. Eggs are occasionally seen with one or two splashes or blotches of very pale brown. It must be remembered, however, that these marked eggs are very exceptional, and represent a selection from vast numbers of the normal type, which is colourless or nearly so. Average measurement, 3·7 inches in length by 2·8 inches in breadth. Incubation is performed chiefly by the female, but the duration of the period is unknown.

DIAGNOSTIC CHARACTERS: The eggs of the Griffon Vulture may be generally identified by the absence of colouring matter; the marked varieties, however, cannot safely be distinguished from those of the Black Vulture. It may be remarked that the latter species always nests in a tree and the Griffon always on rocks.

Family VULTURIDÆ. Genus NEOPHRON.

EGYPTIAN VULTURE.

NEOPHRON PERCNOPTERUS (*Linnæus*).

(British: Very rare abnormal autumn migrant.)

Single Brooded. Laying season, April and May.

BREEDING AREA: South-western Palæarctic region and North-eastern Ethiopian region. The Egyptian Vulture breeds in the mountains of the Spanish Peninsula, the Pyrenees, the Alps, and the mountainous regions of Southern Europe generally, especially those of Turkey, Greece, and Asia Minor. Eastwards it breeds in the Caucasus, Northern Persia, and Turkestan. South of the Mediterranean it breeds in the Canaries, Madeira and Cape Verd Islands, and across Africa north of the Great Desert from Morocco to Egypt, and southwards to Nubia.

BREEDING HABITS: The Egyptian Vulture is only a summer visitor to the countries north of the Mediterranean, reaching its old breeding places early in March in the west, but not until late in that month in the extreme east, and in Asia Minor. It is not gregarious during the nesting season, breeding in scattered pairs. There is little doubt that it pairs for life, and appears generally to return to one particular spot to rear its young. The favourite breeding places of this Vulture are ranges of limestone cliffs, and as the bird appears to be less fastidious in the selection of a site, its nests are more generally distributed than those of the Griffon Vulture, and as a rule much more accessible. In Turkey, however, the Egyptian Vulture often nests in a cypress tree or on the walls of a mosque, and it has even been known to use the old nest of a Short-toed Eagle situated in a cork oak: old nests of the Bearded Vulture are also employed. In Algeria the old nests of Ravens are frequently tenanted. This Vulture, when it does make its own nest, is not much of an architect, usually contenting itself with a few dead sticks—a mere platform—on which a little dry grass and wool has been carelessly arranged. The nest and its immediate vicinity are splashed with the white droppings of the parent birds. This Vulture when disturbed contents itself with wheeling round about the spot or occasionally sweeping past the nest, never showing any pugnacious tendency.

RANGE OF EGG COLOURATION AND MEASUREMENT: The eggs of the Egyptian Vulture are usually two in number, but sometimes three are found. They are very handsome objects, varying in ground colour from buffish-white to yellowish-white, washed, clouded, blotched, and spotted with rich brownish-red. On some eggs the colour is so thickly if somewhat unevenly washed over the entire surface that no ground colour is visible; on others the markings take the form of blotches and spots amongst which the ground colour shows plainly enough, except on the larger end of the egg, where the blotches become confluent and form a zone. Occasionally this zone is round the smaller end of the egg. Average measurement, 2·6 inches in length by 2·0 inches in breadth. Incubation is performed chiefly by the female, but the duration of the period is unknown.

DIAGNOSTIC CHARACTERS: The rich colour and size of the eggs of this Vulture prevent them from being confused with those of any other European species.

Family FALCONIDÆ. Genus HIEROFALCO.
Sub-family FALCONINÆ.

WHITE JER-FALCON.

HIEROFALCO CANDICANS (*Gmelin*).

(British: Rare nomadic winter migrant.)

Single Brooded. Laying season, May and June.

BREEDING AREA: Northern Nearctic region. The White Jer-Falcon is only known to breed in North Greenland, where it has been noticed nesting as far north as Grinnell Land in lat. 79° 41', and Arctic North America westwards apparently to Alaska and the Asiatic coast of Bering Strait.

BREEDING HABITS: But little is known of the nidification of this handsome bird, and its habits during the breeding season remain for the most part undescribed.

The breeding haunts of the White Jer-Falcon are principally confined to woodlands, to ranges of cliffs near the sea, and possibly to the ridges and steep banks of the barren grounds of Arctic America. Of the pairing habits of this bird nothing is known. It is not in any way gregarious, and appears to live a more or less solitary life except when the necessities of the breeding season compel a closer companionship between the sexes. The nest of this Falcon, often made whilst deep snow is still upon the ground, is a mere hollow on a ledge or shelf of some cliff; or it is said the bird sometimes takes possession of the deserted nest of some other species, usually one at the top of a pine or other tree. Whether it ever makes a nest for itself is not clear: Macfarlane infers that it does. It has also been known to nest on the rough ground at the side of a steep hill. It is very noisy, pugnacious, and daring when disturbed at the nest; Sir John Richardson gives a graphic account of the actions of a pair of these Falcons that resented his interference with their nest, which was built on a lofty precipice on the shores of Point Lake in lat. $65\frac{1}{2}°$.

RANGE OF EGG COLOURATION AND MEASUREMENT: The eggs of the White Jer-Falcon are three or four in number. They are creamy-white in ground colour, but so closely freckled, clouded, blotched, and washed with surface markings, that but little of this is ever visible. The surface markings, are orange-brown, brick-red, and dark reddish-brown in colour; the underlying ones clear lilac-gray. The usual type has the surface colour more or less evenly washed over the entire surface; in another type the colour is broken up into blotches and spots, many of them confluent, and most numerous on the larger end of the egg; whilst a third and much rarer type has most of the spots gray underlying ones with an occasional pale brown surface mark here and there.

Average measurement, 2·3 inches in length by 1·8 inch in breadth. Incubation (often begun as soon as the first egg is laid) is apparently performed chiefly by the female, but the duration of the period is unknown.

DIAGNOSTIC CHARACTERS : I know of no character which will serve to distinguish the eggs of this Falcon from those of the other Jer-Falcons. As a rule the locality is of the first importance in their identification. From eggs of the Peregrine they may be distinguished by their larger size, more ovate form, coarser grain, and on an average orange-brown instead of reddish-brown colour.

Family FALCONIDÆ.   Genus HIEROFALCO.
Sub-family FALCONINÆ.

ICELAND JER-FALCON.

HIEROFALCO ISLANDUS (*Gmelin*).

(British : Rare nomadic winter migrant.)

Single Brooded.   Laying season, middle of April to middle of May.

BREEDING AREA : Extreme North-western Palæarctic region. The Iceland Jer-Falcon appears only to breed in Iceland.

BREEDING HABITS : Although great numbers of eggs of the Iceland Jer-Falcon have been collected, the nesting habits of the bird itself are little known. Its favourite and apparently only nesting places are on the cliffs and crags that bound the sea-coast, or that hem in the inland lakes and fjords. This Falcon is not known to make any nest, beyond a mere hollow in the scanty soil on a ledge or shelf of the cliffs, but it is said occasionally to make use of the deserted nest of a Raven in these localities. The bird probably pairs for life, and either uses the same nesting site yearly, or has a selection of several in various parts of its rocky haunt which are used alternately. Of its actions at the nest nothing of exceptional interest has been recorded.

RANGE OF EGG COLOURATION AND MEASUREMENT: The eggs of the Iceland Jer-Falcon are three or four in number, but in some cases apparently only two. They are exceedingly handsome, and subject to much the same range of variation as those of the Greenland Jer-Falcon. Eggs in the same clutch not unfrequently vary considerably amongst themselves, as is often the case in this group of birds. They are buffish-white or very pale buffish-brown in ground colour, spotted, blotched, mottled, and washed with reddish-brown, brick-red, and paler brown, and with a few and indistinct underlying markings of gray. The usual type is so thickly, if somewhat unevenly, washed with reddish-brown as to conceal all trace of the ground colour, but other types occur in which the markings — either brick-red or reddish-brown—are scattered and defined, and show much of the pale ground between them. Average measurement, 2·4 inches in length by 1·9 inch in breadth. Incubation is probably performed chiefly by the female, but the duration of the period is unknown.

DIAGNOSTIC CHARACTERS: The eggs of the Iceland Jer-Falcon present no constant character by which they can be distinguished from those of the other Jer-Falcons. The locality, in this case, is sufficient to identify them. Their larger size, coarser grain, and generally more ovate form prevent them from being confused with those of the Peregrine.

Family FALCONIDÆ. Genus HIEROFALCO.
Sub-family FALCONINÆ.

SCANDINAVIAN JER-FALCON.

HIEROFALCO GYRFALCO (*Linnæus*).

(British : Rare nomadic winter migrant.)

Single Brooded. Laying season, April and May.

BREEDING AREA : North-western Palæarctic region, so far as is at present known. The Scandinavian Jer-Falcon breeds on the fells of Norway and Sweden, especially in the Norwegian province of Finmark. Whether this form of Jer-Falcon inhabits the northern portions of Russia and Siberia seems by no means clear. It may ultimately be found to do so, and to interbreed with *H. candicans* in the far east, as it appears to do with *H. islandus* in the west.

BREEDING HABITS : With the exception of the observations made by Wolley and the Swedish naturalist Collett, we have little information concerning the nesting habits of the Scandinavian Jer-Falcon. Even within the limits of this restricted observation there is much evidence of a conflicting nature, which seems to suggest that the bird is well able to adapt itself to local circumstances. Whether this Falcon ever builds an elaborate nest for itself is by no means clear. Collet says that the bird almost always builds a nest at the top of a large fir tree ; and an egg was brought to Wolley with the information that it had been taken from a nest in a tree. On the other hand, Wolley's personal experience was to find the bird breeding on a ledge of the rocks, he describing the nests as very flat and large, made of dead, barkless sticks, and lined with dry grass, or composed of fresh sticks, lined with willow twigs and sedge ;

other nests were lined with feathers. There is no evidence to prove that the Falcons made these nests themselves, and whether they were deserted nests of Ravens or other birds still seems to me a moot point. When disturbed at the nest the old birds become very anxious, often careering wildly about uttering shrill chattering cries.

RANGE OF EGG COLOURATION AND MEASUREMENT: The eggs of the Scandinavian Jer-Falcon are three or four in number. They are very handsome, and present practically the same characteristics, the same range of variation as those of the two other Jer-Falcons already described. The ground colour varies from creamy-white to buffish or reddish-white, but as regards the latter tint it is difficult to say whether a pale wash of surface colour has not produced it. The surface spots and blotches are reddish-brown and orange-brown of various shades, the underlying markings are pale gray. The usual type is more or less evenly washed with colour, with darker patches here and there; less frequent types have the markings defined and of varying size, or the gray shell markings clear, numerous, and large. Average measurement, 2·3 inches in length by 1·8 inch in breadth. Incubation is chiefly performed by the female, but the duration of the period is unknown.

DIAGNOSTIC CHARACTERS: The locality is the best guide to the identification of the eggs of this Jer-Falcon. See also remarks on pp. 125, 126.

Family FALCONIDÆ.   Genus FALCO.
Sub-family FALCONINÆ.

ORANGE-LEGGED HOBBY.

FALCO VESPERTINUS, *Linnæus*.

(British : Rare abnormal spring and autumn migrant.)
Single Brooded. Laying season, May and June.

BREEDING AREA: North-western Palæarctic region. The Orange-legged Hobby breeds throughout Russia, south of about lat. 65°, in the Danubian provinces, and in Hungary. Eastwards it breeds in South-western Siberia, at least as far east as the valley of the Yenesay.

BREEDING HABITS: The Orange-legged Hobby reaches its European breeding haunts during the last half of April in the west, but is a week or so earlier in the extreme south and east. It is a gregarious species on passage, and to a great extent during the breeding season too, the extent of the colonies depending a good deal on the amount of accommodation available. The principal breeding haunts of this Falcon are well-wooded localities, especially parks, swamps covered with scattered trees, pleasure-grounds, and large gardens. The Orange-legged Hobby apparently never makes its own nest, but selects the deserted one of a Crow, a Magpie, or a Rook, in which to deposit its eggs. In rookeries it may be said to breed in colonies, but elsewhere it lives in scattered pairs simply because the nests it breeds in are isolated. In a rookery as many as five or six nests are tenanted in a single tree. I cannot find that these selected nests undergo any alteration, but the lining is probably removed. Of the habits of this Falcon at the nest and during the pairing and incubating periods nothing appears to have been recorded. The bird probably pairs for life, and seems to visit certain places annually.

K

RANGE OF EGG COLOURATION AND MEASUREMENT: The eggs of the Orange-legged Hobby are from four to six in number. They are rotund in shape, creamy-white in ground colour, washed, spotted, and blotched with orange-brown. As a rule most of the surface markings are confluent, and more or less evenly washed over the entire surface, concealing every trace of ground colour; but varieties may be seen in which the markings are more broken and disconnected. The eggs of this Falcon cover almost precisely the same range of variation as those of the Common Kestrel. Average measurement, 1·5 inch in length by 1·2 inch in breadth. Incubation, performed chiefly by the female, lasts about a month.

DIAGNOSTIC CHARACTERS: I know of no character by which the eggs of the Orange-legged Hobby may be distinguished from those of the Lesser Kestrel. They are generally smaller than those of the Common Kestrel and perceptibly yellower in tint. Goebel also remarks that the grain is finer, and the weight of the empty shell proportionately and absolutely less than that of the eggs of the Common Kestrel. They require, however, the most careful identification.

Family FALCONIDÆ.　　　　　　　　Genus FALCO.
Sub-family FALCONINÆ.

LESSER KESTREL.

FALCO CENCHRIS, *Naumann.*

(British : Very rare abnormal spring and autumn migrant.)

Single Brooded. Laying season, end of April and in May.

BREEDING AREA: South-western Palæarctic and North-eastern Ethiopian region. The Lesser Kestrel breeds in the Spanish Peninsula, in Sardinia and Sicily, in Southern Austria, in Turkey and Greece, and in the extreme south of Russia. Eastwards it breeds in Asia Minor, Palestine, the Caucasus, Persia, and Western Turkestan. South of the Mediterranean it breeds in Northern Africa from Morocco to Egypt.

BREEDING HABITS: The Lesser Kestrel is a bird of passage, and reaches its breeding grounds from its winter quarters in Southern Africa during the last half of March. No raptorial bird is more gregarious, perhaps, than the Lesser Kestrel; it migrates in flocks, and breeds in colonies of varying size. Its favourite breeding haunts are the vicinity of ruins and rocky country fairly well timbered. It is also extremely partial to villages and small towns. It is probable that this Kestrel pairs for life, and returns season by season to the same nesting places. Like its ally the Common Kestrel, it never makes a nest for itself, but selects some hole in a rock, a building, or a tree. The bird is very partial to holes in ruins, church towers, and even the caves of houses. Colonies of this species have been observed even in the streets of a town under the eaves; a deserted nest in a tree is sometimes selected. The eggs are laid in slight hollows, and are generally surrounded with pellets and food refuse. The bird is a close sitter, and views disturbance with little concern, the members of a colony flying to and fro and in and out of their nest-holes with little shyness or fear.

RANGE OF EGG COLOURATION AND MEASUREMENT: The eggs of the Lesser Kestrel are from four to seven in number, five or six being the usual clutch. They are globular in form, and yellowish-white in ground colour, but very often this is so washed with surface colour as to appear pale brick-red. The markings are pale and dark orange-brown, and are generally washed over the entire surface concealing the ground colour, but varieties occur in which the spots and blotches are scattered, and the pale ground shows distinctly between them. The eggs of this species cover much the same range of variation as those of the Common Kestrel and the Orange-legged Hobby. Average measurement, 1·4 inch in length by 1·1 inch in breadth. Incubation is performed by both sexes, but the duration of the period is unknown.

DIAGNOSTIC CHARACTERS: There is no constant character by which the eggs of the Lesser Kestrel can be distinguished from those of the Orange-legged Hobby. They are on an average smaller. From eggs of the Common Kestrel their small size and yellower tints—orange not red—serve to distinguish them. They should always be carefully identified.

Family FALCONIDÆ. Genus AQUILA.
Sub-family *AQUILINÆ*.

SPOTTED EAGLE.

AQUILA NÆVIA, *Meyer*.

(British : Very rare abnormal winter migrant.)

Single Brooded. Laying season, May.

BREEDING AREA: Western Palæarctic region. The small race of Spotted Eagle has a somewhat restricted breeding range, being confined to Prussia and the North German Confederation, the Baltic Provinces of Russia, Poland, and South-west Russia down the valleys of the Dnieper and the Dniester to the Caucasus.

BREEDING HABITS: The Spotted Eagle reaches the north-western limits of its breeding range at the beginning of April. Its breeding haunts are the great forests, especially such as are swampy or situated near to morasses and bogs. This Eagle is not gregarious, but lives in scattered pairs. There can be little doubt that it pairs for life, and yearly returns to one particular spot to breed, generally making a new nest in the vicinity of the one of the previous season. The nest is placed at varying heights of from thirty to eighty feet from the ground on large trees—beeches, firs, oaks, and birches being used indiscriminately—either near the top where several branches meet, or on a limb near the trunk. A nest is recorded as having been built upon the ground. It is a large flat structure, several feet across, and sometimes a couple of feet in height, composed externally of sticks of various thickness, the stouter ones being placed at the bottom, and lined with fresh green twigs with the leaves attached, or green grass. The bird sits closely, but when disturbed makes little demonstration, either flying completely away, or retiring to a distant tree to watch the fate of its nest.

RANGE OF EGG COLOURATION AND MEASUREMENT: The eggs of the Spotted Eagle are generally two in number, very often only one, and very exceptionally as many as three. The Seebohm collection contains a very handsome series of these eggs from the collection of Dr. Holland. They vary from grayish-white to creamy-white in ground colour, blotched and spotted with reddish-brown, brick-red, and rich dark blood-red, and with underlying markings of pale purplish-brown. They vary considerably in the amount and intensity of the markings, some being very sparsely spotted, others richly and boldly blotched, many of the patches being confluent, either at the large or small end of the egg, where they are also most numerous. A rare type has most of the markings underlying ones and very large. Average measurement, 2·5 inches in length by 2·1 inches in breadth. Incubation is performed chiefly by the female, and is said to last three weeks.

DIAGNOSTIC CHARACTERS: The size is the best character to distinguish the eggs of this Eagle, at least as far as European species are concerned. The locality too is of the greatest importance, otherwise they may be confused with the eggs of *Aquila clanga*.

Family FALCONIDÆ. Genus MILVUS.
Sub-family BUTEONINÆ.

BLACK KITE.

MILVUS ATER (*Gmelin*).

(British : Very rare abnormal spring migrant.)

Single Brooded. Laying season, April, May, and June.

BREEDING AREA: South-western Palæarctic region. The typical form of the Black Kite breeds throughout Europe in suitable localities from Finland and Central Russia southwards to the Caspian, Black, and Mediterranean Seas, with the exception of Scandinavia, Denmark, the Netherlands, Belgium, and Northern France. Eastwards it breeds in Asia Minor, Palestine, Persia, and Turkestan, while south of the Mediterranean it is widely dispersed in North-west Africa north of the Atlas.

BREEDING HABITS: The Black Kite reaches its European breeding grounds in March or early in April. Its favourite haunts are marshy forests, but it may also be met with in a great variety of other places. In Algeria I met with it on the bare mountains, and on desolate stony plains, as well as in towns. It is also known to breed in or near various European towns. Few other birds of prey are more widely dispersed or inhabit such a varied description of scenery. The Black Kite is a gregarious bird, and certainly breeds in colonies in many places where it is not much molested by man. Mr. Saunders met with a colony containing more than ten nests in a small wood in Spain ; Captain Verner found great numbers of nests in the pine woods on the north bank of the Guadalquivir ; whilst in Algeria I remarked a colony in the stupendous gorge of El Kantara. The bird probably pairs for life. The nest of this Kite is built in a variety of situations depending a good deal upon the kind of accommodation afforded. In some districts trees are the favourite situation, sometimes very low ones growing amongst reeds, as recorded by Bogdanow, in the delta of the Volga, and sometimes amongst the roots of trees growing out of cliffs, as recorded by Mr. Salvin in Algeria, in the Eastern Atlas. In other districts a ledge or shelf or crevice of a cliff is selected; in others a ruin or a tower. The nest varies somewhat in size, the smallest apparently being made in Southern Russia, where Goebel states that the head and tail of the sitting bird project over each side of the structure. Larger nests measure a yard or more across. The nest is rather flat, and composed externally of sticks, and lined with dry moss, paper, and rags, usually in a more or less filthy condition. Captain Verner also found a mixture of dry dung as well as paper in the lining. Coloured rags and papers are frequently preferred. When disturbed from the nest the old birds fly round and round above the spot uttering shrill, tremulous cries of alarm.

RANGE OF EGG COLOURATION AND MEASUREMENT: The eggs of the Black Kite are from two to five in number, but the latter is very exceptional. In Prussia and Spain two is the normal clutch, in Southern Russia three. They vary in ground colour from white to the palest perceptible blue, spotted and blotched with rich reddish-brown and pale brown, and with underlying markings of lilac-gray. They are subject to much variation both in the intensity and the character of the markings. On some eggs large blotches and splashes of dark brown occur intermingled with spots and streaks of lighter brown; on others the markings are very minute and dusted over the entire surface, most numerous round the larger end; others are clouded and washed with pale brown, and sparingly marked with richer brown. Rarer types have most of the markings gray underlying ones, or the brown surface markings distributed in net-like and confluent streaks. Average measurement, 2·2 inches in length by 1·7 inch in breadth. Incubation, performed chiefly by the female, lasts about three weeks. If the first clutch of eggs be taken others are generally laid.

DIAGNOSTIC CHARACTERS: The eggs of the Black Kite cannot be distinguished from those of the Common Buzzard, the Common Kite, and the Rough-legged Buzzard. They require careful identification. The rag-lined nest is characteristic of the two Kites only, but in every case great care should be used in identifying them.

Family FALCONIDÆ. Genus ELANOIDES.
Sub-family BUTEONINÆ.

SWALLOW-TAILED KITE.

ELANOIDES FURCATUS (*Linnæus*).

(British : Very rare abnormal autumn migrant.)

Single Brooded. Laying season, May.

BREEDING AREA: Southern Nearctic region and extreme north of Neotropical region. The Swallow-tailed Kite breeds from the mountains of Central America northwards into the United States as far as Southern Wisconsin, and east of the Rocky Mountains.

BREEDING HABITS: The Swallow-tailed Kite is a regular migrant to the United States, reaching its breeding areas early in April. It is a decidedly gregarious bird whilst on passage, and so far as is known apparently breeds in societies. Of its pairing habits nothing definite has been observed beyond what was recorded by Audubon. Although this graceful bird is so common in certain districts, but little has been recorded of its nidification or of its habits during the season of reproduction. According to Audubon, this Kite pairs directly after its arrival, and the courtship is carried on in mid-air like that of the Swift. The nest is described by this naturalist as being made in the highest branches of lofty trees growing on the banks of a pond or river, and resembling that of a crow, composed externally of sticks intermixed with moss and lined with coarse grass and feathers. No recent observer appears to have described the nest of this bird from personal observation. Mr. Dresser states that in Texas he was assured that this species nested in oak, cotton-wood, and sycamore trees.

RANGE OF EGG COLOURATION AND MEASUREMENT: The eggs of the Swallow-tailed Kite are said by Audubon to be from four to six in number, but other observers assert that two is the regular clutch. I have only examined two eggs of this species, and they are, or were, in the collection of Mr. Dresser and were taken in Iowa. Mr. Raine (*op. cit.*, pl. iv. p. 130) figures an egg of this species, very richly marked on the larger end. They are the palest of blue—almost white—in ground colour, spotted and blotched with deep reddish-brown. One of Mr. Dresser's eggs is handsomely blotched over most of the surface; the other is sparingly and minutely speckled, with only one or two irregular blotches. Average measurement of three specimens, 1·9 inch in length by 1·5 inch in breadth. Incubation period unknown.

DIAGNOSTIC CHARACTERS: The bluish ground colour and bold blotches, combined with the size, are sufficient to determine the eggs of this species. They closely resemble those of the Sparrow-Hawk or even the Sharp-shinned Hawk in colour, but are much larger.

Family FALCONIDÆ.   Genus ARCHIBUTEO.
Sub-family BUTEONINÆ.

ROUGH-LEGGED BUZZARD.

ARCHIBUTEO LAGOPUS (*Gmelin ex Brünnich*).

(British : Fairly common spring and autumn coasting migrant.)

Single Brooded. Laying season, May and early June.

BREEDING AREA: Northern Palæarctic region. The Rough-legged Buzzard breeds in all parts of Scandinavia suited to its requirements as far as the North Cape, and North Russia as low as the Baltic Provinces, where, however, it is rare. Eastwards it breeds in Northern Siberia to at least as far east as the watershed of the Yenesay and the Lena.

BREEDING HABITS : The Rough-legged Buzzard returns from the south to its northern breeding grounds in April. Although the bird cannot perhaps be classed as gregarious whilst on passage, it is by no means an unsociable one, and even in the breeding season many pairs may be found nesting within a comparatively small area. The favourite breeding grounds of this Buzzard are rocky fells and hilly tracts of country in which the forests are more or less broken up and studded with swamps and sheets of open water. It shows less partiality for the interior of forests than the Common Buzzard, and is much more of a mountain and open country species. I would suggest that this bird pairs for life, and yearly resorts to one particular spot to breed. The nest is variously placed on rocks, on fell ridges, or on trees, the largest and most elaborate structures being usually in the latter sites. Externally it is made of sticks and twigs, and lined with fine dry grass—a large flat and open structure. Many nests on ridges and shelves of rock are mere hollows lined with a little grass. The bird is a remarkably close sitter, and when disturbed wheels round and round the nesting place, sometimes uttering a mewing note, and usually quickly joined by its mate. Mr. Harvie-Brown gives a very interesting instance of a male bird getting another mate, after the first female had been shot from the nest, within twenty-four hours.

RANGE OF EGG COLOURATION AND MEASUREMENT: The eggs of the Rough-legged Buzzard are from two to five in number, three or four being an average clutch. They vary from white suffused with pale brown to pale bluish-green in ground colour, blotched, clouded, spotted, and streaked with rich reddish-brown and pale brown, and with underlying markings of violet-gray. They are subject to considerable variation, the amount of the spotting and the intensity of its colour differing considerably even in the same clutch. The two rarest types are perhaps those in which the markings are small and delicately streaked or pencilled over most of the surface, and in which most of the markings are underlying ones. The eggs of this bird cover the same range of variation as those of the Common Buzzard. Average measurement, 2·25 inches in length by 1·8 inch in breadth. Incubation is performed almost if not entirely by the female, and lasts about three weeks.

DIAGNOSTIC CHARACTERS: The eggs of this species cannot be distinguished from those of the Common Buzzard. As a rule they are more heavily marked and a trifle larger. The breeding range is also much more Arctic.

Family FALCONIDÆ. Genus ASTUR.
Sub-family ACCIPITRINÆ.

GOSHAWK.

ASTUR PALUMBARIUS (*Linnæus*).

(British : Possibly bred ; rare abnormal spring and autumn migrant.)

Single Brooded. Laying season, latter half of April and first half of May.

BREEDING AREA: Palæarctic region. The Goshawk breeds locally and somewhat sparingly throughout Europe in the forest districts from the Mediterranean northwards to the limits of trees. South of the Mediterranean it has been known to breed in Morocco. Eastwards it breeds throughout Siberia south of the limit of forest growth, and in Asia Minor, Northern Palestine, Persia, Turkestan, the Himalayas, Mongolia, and Northern China.

BREEDING HABITS: The Goshawk is a migrant only in the most northerly portions of its area. Its favourite breeding haunts are forests, large woods, and plantations ; although the bird hunts for prey a good deal in open country it invariably nests within the cover of trees. I do not find that this species shows any gregarious tendency. It lives in solitary isolated pairs, and unquestionably mates for life, breeding in the same nest year after year, or making a new nest in the old neighbourhood. The nest is generally placed high up in a lofty tree, either in a crotch formed by several forking branches, or on a broad horizontal limb close to the trunk. According to the age of the nest it varies somewhat in size, the biggest structures being those that have been increased in bulk during many successive seasons. It is broad and flat, but the cup containing the eggs is somewhat deep. Externally it is made of sticks and twigs and lined with finer twigs, roots, moss, and flakes of bark. The bird is a close sitter, but when flushed usually flies away making little or no demonstration.

RANGE OF EGG COLOURATION AND MEASUREMENT: The eggs of the Goshawk are from three to five in number, four being the average and usual clutch. They are pale bluish-green, and generally entirely devoid of markings, except perhaps a few stains of yellowish-brown. Very rarely eggs are seen with a few rich brown spots: Wolley states that occasionally they are marked with pale olive. Average measurement, 2·25 inches in length by 1·75 inch in breadth. Incubation, performed chiefly if not entirely by the female, is said to last three weeks.

DIAGNOSTIC CHARACTERS: The eggs of the Goshawk may be generally distinguished by their size and absence of markings, but as unspotted eggs of some of the Buzzards are the same size, they require careful identification.

Family FALCONIDÆ.   Genus ASTUR.
Sub-family ACCIPITRINÆ.

AMERICAN GOSHAWK.

ASTUR ATRICAPILLUS (*Wilson*).

(British: Very rare abnormal autumn migrant.)

Single Brooded. Laying season, May.

BREEDING HABITS: Northern Nearctic region. The American Goshawk breeds in some of the northern

States, and throughout British North America and Alaska, up to the limits of forest growth.

BREEDING HABITS : The American Goshawk is only a migrant from the colder portions of its breeding area. Its favourite breeding haunts are in pine and other woods, bluffs (small clumps of trees on prairies), and the outskirts of forests. In its habits it is not known to differ from its Old World ally, from which it is possibly only sub-specifically distinct. The flat open nest is generally made in a fir tree, on a horizontal branch close to the trunk, and is made externally of large and small sticks, and lined with dead leaves and flakes of bark.

RANGE OF EGG COLOURATION AND MEASUREMENT : The eggs of the American Goshawk are from three to five in number, three or four being generally found, and sometimes only two. They are pale bluish-green, and, so far as I can determine, never show any traces of spots beyond an occasional nest stain. Average measurement, 2·2 inches in length, by 1·7 inch in breadth. Short eggs are generally the broadest, and *vice versâ*. Incubation is performed chiefly by the female, but the duration of the period is undetermined. It is probably the same as that of the Palæarctic species.

DIAGNOSTIC CHARACTERS: The eggs of the American Goshawk cannot be distinguished from those of its Old World ally, nor from certain and unspotted varieties of those of some other species.

Family ANATIDÆ.	Genus CYGNUS.
Sub-family CYGNINÆ.

HOOPER SWAN.

CYGNUS MUSICUS, *Bechstein.*

(British : Common autumn and winter migrant.)

Single Brooded. Laying season, late in May and in June.

BREEDING AREA: Northern Palæarctic region. The Hooper Swan breeds in the Arctic regions of Europe and Asia, from the Atlantic probably to the Pacific. It breeds in Iceland and in Scandinavia north of the Arctic circle, but in Finland and Northern Russia is said to do so as low as lat. 62°. Eastwards it breeds across the tundras of Arctic Siberia as far as Bering Strait.

BREEDING HABITS: The Hooper Swan begins to appear in its summer haunts about the middle of May, just as the ice on the great rivers that flow into the Arctic seas is breaking up, and it continues to arrive in vast numbers up to the beginning of June, following the break-up of winter northwards. This Swan is not gregarious during the breeding season, living in scattered pairs, each pair keeping its own haunt free from intrusion. The Hooper Swan mates for life. Its favourite breeding grounds are situated on the islands in the deltas of the Arctic rivers, or near the lakes on the open tundras, or on the banks of the creeks that run inland from the main river. The nest is generally made amongst willow scrub, or the tall grasses and reeds that fringe the pools. It is a huge pile of coarse grass, sedge, and fragments of herbage, often largely increased in bulk as incubation advances—elevated and strengthened, as it were, in anticipation of any sudden rise of the adjoining water.

RANGE OF EGG COLOURATION AND MEASUREMENT: The eggs of the Hooper Swan are from four to seven in number. In Northern Russia two to four are described as the usual clutch; in Iceland five; whilst the larger clutches are probably laid by females in the prime of life. They are creamy-white, rough in grain, and nearly oval in shape. Average measurement, 4·5 inches in length by 2·8 inches in breadth. Incubation, performed largely if not entirely by the female, lasts from five to six weeks.

DIAGNOSTIC CHARACTERS: The only eggs likely to be confused with those of the present species are the eggs of Bewick's Swan, but the latter are always smaller in size (either in length or breadth or both), and the empty shells much lighter in weight.

Family ANATIDÆ.
Sub-family *CYGNINÆ*.

Genus CYGNUS.

BEWICK'S SWAN.

CYGNUS BEWICKI, *Yarrell*.

(British : Fairly common autumn and winter migrant.)

Single Brooded. Laying season, June.

BREEDING AREA: Northern Palæarctic region. Bewick's Swan breeds on the tundras of Arctic Europe and Asia, probably as far east as Bering Strait. It appears to breed in the valleys of the Petchora (where Messrs. Seebohm and Harvie-Brown obtained the first authenticated eggs), the Obb, the Yenesay, and the Lena, as well as on the islands of the Arctic Ocean north of these areas.

BREEDING HABITS: Bewick's Swan reaches its Arctic breeding grounds towards the end of May or early in June, as soon as the south wind has sufficiently ensured the presence of open water. Although gregarious on migration, the bands disperse for the breeding season and nest in scattered pairs. There can be little doubt that this Swan mates for life, but builds a new nest each season. Its breeding grounds are not known to differ from those of the Hooper Swan—the shores of lakes on the tundra, the banks of creeks, and scrub-clothed islands in the deltas of the Arctic rivers. But little is known of the habits and nest of this Swan, and its eggs, so far as I know, have never been taken from the nest by any scientific collector or naturalist. Eggs of this bird, of whose authenticity there can be no reasonable doubt, were brought to Messrs. Harvie-Brown and Seebohm during their visit to the Petchora in 1875, which had been obtained on the island of Pyonin in the delta of that river; whilst others were secured under similar circumstances by Mr. Seebohm during his sojourn in the Yenesay in 1877, taken from an island and from the mainland near that river's delta. Natives describe the nest as being like that of the Hooper Swan, and built in a similar situation. This species is remarkably wary at the nest.

RANGE OF EGG COLOURATION AND MEASUREMENT: The full number of eggs in a clutch of this Swan is not known with certainty, but more than three have not yet been found in any one nest. They are almost pure white. Average measurement, 4·0 inches in length by 2·6 inches in breadth. The duration of the period of incubation is unknown. Probably the female undertakes most of, if not entirely all the duty.

DIAGNOSTIC CHARACTERS: The eggs of this Swan may be distinguished from those of the Hooper Swan—

the only species with which they are likely to be confused—by their smaller size (either in length or breadth or both) and lighter shell weight. They are also perceptibly whiter.

Family ANATIDÆ.
Sub-family ANSERINÆ.

Genus CHEN.

LESSER SNOW GOOSE.

CHEN HYPERBOREUS (*Pallas*).

(British : Very rare abnormal autumn migrant.)

Single Brooded. Laying season, June.

BREEDING AREA : Apparently North-western Nearctic region, and probably extreme North-eastern Palæarctic region. The Lesser Snow Goose is only known to breed in the north-west of Arctic America (Esquimaux Lake, Liverpool Bay, etc.), in Alaska, and possibly on the tundras of the Tchuski Land in North-eastern Siberia, and the islands off that coast, as the bird was met with by Pallas in that area.

BREEDING HABITS : The Lesser Snow Goose reaches its breeding grounds towards the end of May or early in June. It migrates in flocks, but whether it is gregarious during the breeding season is not known. It is probably to a certain extent social during this period, and seems to pair for life. Only the most meagre details have been recorded of this bird's habits during the season of reproduction. Nests found on an island near the mouth of the Anderson river, are described by MacFarlane as mere hollows in the sandy ground warmly lined with down, but he does not appear to have seen them *in situ*.

RANGE OF EGG COLOURATION AND MEASUREMENT: The eggs of the Lesser Snow Goose are said usually to be five in number. They are dull white, with little or no polish. Average measurement, 3·4 inches in length by 2·2 inches in depth. Incubation is performed by the female, but the duration of the period is unknown.

DIAGNOSTIC CHARACTERS: I know of no character which will serve to distinguish the eggs of the Lesser Snow Goose from those of several allied species—notably from those of the Gray Lag Goose (*Anser cinereus*) and the Snow Goose (*Chen hyperboreus nivalis*). Information is much to be desired respecting these eggs. So few specimens are at present known that it is not safe to generalize on their characteristics.

Family ANATIDÆ.　　　　　　　　　　　　Genus ANSER.
Sub-family *ANSERINÆ*.

BEAN GOOSE.

ANSER SEGETUM (*Gmelin*).

(British: Common autumn and winter migrant.)

Single Brooded. Laying season, June.

BREEDING AREA: Northern and Central Palæarctic region. The Bean Goose breeds in Scandinavia north of lat. 64°, but in Northern Russia north of about lat. 65°, above the limits of forest growth. It may probably breed on Nova Zembla, and eastwards is known to do so across Siberia on the tundras, and in a similar climate at high elevations on the mountains in the Baikal area and on the Stanavoi mountains.

BREEDING HABITS: The Bean Goose returns to its

breeding grounds in the Arctic regions as soon as they are free from ice, following the breaking-up rivers northwards in May. The favourite haunts of this Goose are tundras and treeless plains, especially such as are studded with pools and traversed by rivers. The Bean Goose is gregarious and social enough whilst on passage, but appears to breed in scattered pairs. It probably mates for life, as so many other Geese are known to do. The nest is either placed amongst the tall grasses and sedges by the margin of a pool, in the long vegetation clothing an island, or on a hillock on the shore. It is merely a hollow lined with dry grass and other vegetable fragments and lined with down and a few feathers from the old bird's body.

RANGE OF EGG COLOURATION AND MEASUREMENT: The eggs of the Bean Goose are three or four in number. They are rather coarse in grain, and creamy-white. Average measurement, 3·2 inches in length by 2·15 inches in breadth. Incubation is performed almost if not entirely by the female, but the duration of the period is unknown.

DIAGNOSTIC CHARACTERS: The eggs of the Bean Goose cannot always be distinguished from those of the Gray Lag Goose in colour or in size, but are lighter in shell weight. I know of no character whatever by which they may be separated from those of the Pink-footed Goose and the White-fronted Goose. It will thus be seen of what vital importance correct identification is.

Family ANATIDÆ. Genus ANSER.
Sub-family *ANSERINÆ*.

PINK-FOOTED GOOSE.

ANSER BRACHYRHYNCHUS, *Baillon*.

(British : Common autumn and winter migrant.)

Single Brooded. Laying season, June.

BREEDING AREA : North-western Palæarctic region. The Pink-footed Goose is only known to breed in Spitzbergen, and may do so in Franz-Josef Land and Iceland.

BREEDING HABITS : The Pink-footed Goose appears upon its breeding grounds in May or early June. But little is known of its nesting economy, and observers who have been fortunate enough to obtain its eggs have neglected to describe the habits of the bird. It probably mates for life, and nests in scattered pairs, the flocks which have spent the winter in company separating during the short period of incubation. The nest of this Goose is said to be made on low rocks near the sea ; but Messrs. Evans and Sturge relate that in Spitzbergen some nests seemed to be made in high cliffs "a mile or two from the sea." No description of the nest appears yet to have been published, but the structure is not known to differ from that of allied species. The male is said to keep constant watch near the nest to warn his mate of the approach of danger, and if necessary to defend her.

RANGE OF EGG COLOURATION AND MEASUREMENT : The eggs of the Pink-footed Goose are four or five in number. They are somewhat smooth in grain, and creamy-white. Average measurement, $3·15$ inches in length by $2·15$ inches in breadth. Incubation, performed by the female, lasts twenty-eight days.

DIAGNOSTIC CHARACTERS: Unfortunately no character can be given by which the eggs of the Pink-footed Goose can be distinguished from those of the Bean Goose and the White-fronted Goose. They may be separated from those of the Gray Lag Goose by their lighter shell weight.

Family ANATIDÆ.   Genus ANSER.
Sub-family *ANSERINÆ*.

WHITE-FRONTED GOOSE.

ANSER ALBIFRONS (*Scopoli*).

(British: Local autumn and winter migrant.)

Single Brooded probably.   Laying season, June.

BREEDING AREA: Northern Palæarctic region. The White-fronted Goose breeds in the Arctic regions of Europe and Asia. It breeds in Iceland, Northern Scandinavia, and Russia, and eastwards through Northern Siberia, probably as far as Bering Strait.

BREEDING HABITS: The White-fronted Goose returns to its Arctic haunts as soon as they become habitable, which is seldom before the end of May or early in June. It is gregarious whilst on passage, but whether it breeds in societies is not known. Von Middendorff met with it breeding in great numbers on the tundras of the Taimyr peninsula, but the probability is the birds were nesting in scattered pairs. This goose also may mate for life. The nests found by Middendorff are described as being built on the hummocks which are so characteristic of the tundras, and mere hollows lined with down. The nests described by Dall in Alaska, as mere depressions in the sand, and by MacFarlane on the Anderson river as warmly lined with dry grass as well as with down and feathers, belonged, of course, to the Nearctic form of this Goose, *Anser albifrons gambeli*, a larger bird with more black on the underparts, but they may be taken as a fair description of the nest of this species in its entirety, as I described it (with the eggs) in my work on *British Game Birds and Wild Fowl*.

RANGE OF EGG COLOURATION AND MEASUREMENT: The eggs of the White-fronted Goose are from five to seven in number; ten have been found by Dall in Alaska, but this refers to the Nearctic race. They are creamy-white. Average measurement, 3·0 inches in length by 2·0 inches in breadth. Incubation is performed probably by the female, but the duration of the period has not been observed.

DIAGNOSTIC CHARACTERS: I know of no character by which the eggs of this Goose may be distinguished from those of several allied species. They require the most careful identification, or are otherwise worthless as scientific specimens.

Family ANATIDÆ.   Genus ANSER.
Sub-family ANSERINÆ.

LESSER WHITE-FRONTED GOOSE.

ANSER ALBIFRONS MINUTUS, *Naumann*.

(British : Rare autumn and winter migrant.)

Single Brooded probably. Laying season, June.

BREEDING AREA: Northern and Western Palæarctic region. The Lesser White-fronted Goose breeds in the northern portions of Scandinavia, across Arctic Russia and Siberia to at least as far east as the valley of the Yenesay.

BREEDING HABITS: The Lesser White-fronted Goose (which by a slip of the pen in my work on *Game Birds and Wild Fowl* I notice I have described as a Northern *Nearctic* instead of *Palæarctic* species) reaches its northern breeding grounds as soon as the ice on the rivers breaks up and the snow melts from the tundras. It is gregarious in winter, but appears to separate into pairs to breed. Of its breeding habits little definite has been recorded, owing probably to its being confused with its larger ally. Indeed I may here take the opportunity of stating that the nesting economy of most of the Geese is very imperfectly known, and a vast amount of work remains to be done by the oologists of the future. It is a hopeless task to attempt to give anything like a full account of the nidification of these birds in the present state of our knowledge. By pointing out our deficiencies I may possibly help to render them perfect. The nest of this Goose is said not to differ from that of allied species. Of the bird's pairing habits nothing is known.

RANGE OF EGG COLOURATION AND MEASUREMENT :

The eggs of the Lesser White-fronted Goose are from five to seven in number. They are creamy-white. Average measurement, 2·9 inches in length by 2·0 inches in breadth. The duration of the period of incubation is unknown, but the female probably alone performs that task.

DIAGNOSTIC CHARACTERS: In the present state of our knowledge it is impossible to attempt to give any character which might be distinctive. Sufficient reliable information is not yet obtainable on which to found any diagnosis of the eggs of this Goose, providing such is really possible.

Family ANATIDÆ. Genus BERNICLA.
Sub-family ANSERINÆ.

BRENT GOOSE.

BERNICLA BRENTA (*Brisson*).

(British : Common autumn and winter migrant.)

Single Brooded. Laying season, late in June.

BREEDING AREA: North-western Palæarctic region. The typical form of the Brent Goose breeds on Spitzbergen, Franz-Josef Land, and Nova Zembla, and possibly along the coasts of Arctic Siberia and the islands off them, but how far to the east is not known.

BREEDING HABITS: The Brent Goose is a migratory bird, and returns to its breeding grounds in the Arctic regions late in May or early in June. When I wrote my account of the nidification of the Brent Goose in the work on *Game Birds and Wild Fowl*, I relied for my information on the observations made by Captain

Feilden near Knot Harbour in Grinnell Land, as applicable to both forms of Brent Goose occurring in the British Islands. His information, of course, refers to the White-bellied form of Brent Goose, and strictly speaking should be confined to that form. In the utter absence of information, however, concerning the typical form, it is perhaps advisable to continue to allow these details to refer to both forms of Brent, and to trust to future observations to fill in the blanks. This Goose apparently pairs for life, and is gregarious enough during its migrations, but whether it continues so through the nesting season is unknown. It certainly appears to be a social species, many birds breeding within a comparatively small area. The nest is made in a hollow in the ground, and composed of dry grass, moss, and other vegetable fragments, and warmly lined with down. The male keeps careful watch near the nest, ready to give the alarm or to defend it from enemies.

RANGE OF EGG COLOURATION AND MEASUREMENT: The eggs of the Brent Goose are four or five in number. They are somewhat smooth in texture, slightly polished, and creamy-white. Average measurement, 2·75 inches in length by 1·85 inch in breadth. Incubation is apparently performed entirely by the female, but the duration of the period is unknown.

DIAGNOSTIC CHARACTERS: The eggs of this Goose cannot be distinguished from those of the Bernacle Goose. They also closely resemble those of the White-fronted Goose, but—size for size—are perceptibly lighter in shell weight.

Family ANATIDÆ. Genus BERNICLA.
Sub-family ANSERINÆ.

WHITE-BELLIED BRENT GOOSE.

BERNICLA BRENTA GLAUCOGASTER (*Brehm*).

(British : Uncommon autumn and winter migrant.)

Single Brooded. Laying season, June.

BREEDING AREA: Northern Nearctic region. The White-bellied Brent Goose breeds across Arctic America on the coasts and islands lying north of lat. 72° from the Arctic Archipelago east to Greenland and north as far as land is known. Captain Feilden met with this Goose breeding near Knot Harbour in lat. 82½°.

BREEDING HABITS: The White-bellied Brent Goose arrived at its breeding grounds at Knot Harbour on the 9th of June, and shortly afterwards the male and female were observed rising to a great height in a spiral course toying with and chasing each other. This bird is gregarious during winter and whilst on passage, and even in the breeding season continues somewhat social, numbers of pairs nesting in close proximity. The nests were either made on the hillsides, between the snow-line and the sea, or were placed on an island beyond the line of open water, separated from the mainland by rough hummocks of snow and ice. The nests were in hollows and composed of grass, moss, and saxifrages, warmly lined with down. The male is very assiduous in keeping guard over the nest, ready to give the alarm to the female or to defend her and it from enemies.

RANGE OF EGG COLOURATION AND MEASUREMENT : The eggs of the White-bellied Brent Goose are four or five in number. They are somewhat fine in grain, rather polished, and creamy-white. Average measurement, 2·75 inches in length by 1·85 inch in breadth. Incubation is performed by the female, but the duration of the period is unknown.

DIAGNOSTIC CHARACTERS: The eggs of this Goose cannot always be distinguished from those of the Bernacle Goose, but from those of the White-fronted Goose, which they also resemble in size and colour, they may be separated by their lighter shell weight—eggs of the same size being of course selected for the test.

Family ANATIDÆ.   Genus BERNICLA.
Sub-family ANSERINÆ.

BERNACLE GOOSE.

BERNICLA LEUCOPSIS (*Bechstein*).

(British: Common autumn and winter migrant.)

Single Brooded probably. Laying season, probably June.

BREEDING AREA: North-western Palæarctic region. The Bernacle Goose has been met with during summer in Greenland, Iceland, Spitzbergen, and Nova Zembla, and has been said to nest on the Loffoden Islands, but its exact breeding grounds remain to be discovered.

BREEDING HABITS: Nothing is known of the breeding habits and the nest of the Bernacle Goose. This species has been said to breed on the Loffoden Islands, and specimens of the eggs, together with a description of the nest, were sent to Collett, but these eggs are certainly too small. These islands are also far south of the usual haunts of this Goose, which probably nests as far north as land occurs. A description of the nest and the habits of the birds during the breeding season still remains to be written.

RANGE OF EGG COLOURATION AND MEASUREMENT (*in confinement*) : The eggs of the Bernacle Goose, judging from those laid by birds in captivity, are somewhat coarse in grain, with no polish, and creamy-white. Average measurement, 2·85 inches in length by 1·95 inch in breadth. Incubation period unrecorded.

DIAGNOSTIC CHARACTERS: The eggs of the Bernacle Goose are on an average slightly larger than those of the Brent Goose, but no absolutely reliable character can be given by which they may be distinguished from those of that species.

Family ANATIDÆ.   Genus BERNICLA.
Sub-family *ANSERINÆ*.

RED-BREASTED GOOSE.

BERNICLA RUFICOLLIS (*Pallas*).

(British : Very rare abnormal autumn and winter migrant.)

Single Brooded probably. Laying season, early in July.

BREEDING AREA: North-central Palæarctic region. The Red-breasted Goose is only known to breed on the tundras above the limits of forest growth in the valleys of the Obb, Yenesay, Piasina, and Boganida rivers.

BREEDING HABITS: Nothing whatever appears to be known of the habits of the Red-breasted Goose during the breeding season. The bird is migratory, leaving its winter quarters in the Caspian basin and retiring to the Arctic regions to breed. It is certainly very gregarious in its winter resorts and whilst on passage; and judging from the few meagre details recorded I should infer a certain amount of sociability during the breeding season, many pairs nesting within a small area. I cannot find that the nest of this Goose has ever been described by a naturalist. Von Middendorff states that he met with this species breeding in some numbers in the valley of the Boganida, and he was the first naturalist to obtain its eggs. Another nest was discovered on one of the islands in the delta of the Yenesay (inadvertently given as the Petchora in my work on *Game Birds and Wild Fowl*) containing two eggs, one of which was broken, as the female was shot upon them, and the other was brought to Mr. Seebohm, then on his visit to that region. This nest was described by its unscientific discoverer as being like that of a Bean Goose, but not so large.

RANGE OF EGG COLOURATION AND MEASUREMENT : The number of eggs laid by the Red-breasted Goose is unknown ; it is to be regretted that Mr. Seebohm did not at least count the broods swimming with their parents, which he saw on the Yenesay, as this would have furnished some clue to the extent of the clutch. They are creamy-white, obscurely marked with green, smooth in grain, and for a Goose egg remarkably fragile. Average measurement, 2·75 inches in length by 1·76 inch in breadth. The duration of the period of incubation is unknown, as is also which parent performs the task.

DIAGNOSTIC CHARACTERS : If the traces of an underlying green shell that show here and there through the creamy-white are constant, then this peculiarity, combined with the very fragile shell, is enough to distinguish them from those of every other Goose known to me. If my memory serves me correctly Middendorff figures the egg of this Goose displaying similar indistinct pale green patches.

Family ANATIDÆ. Genus TADORNA.
Sub-family ANATINÆ.

RUDDY SHELDRAKE.

TADORNA CASARCA (*Linnæus*).

(British : Rare abnormal autumn and winter migrant.)

Single Brooded. Laying season, April and May.

BREEDING AREA: Southern Palæarctic region and North-eastern Ethiopian region. The Ruddy Sheldrake breeds sparingly in the south of Spain and in North Africa from Morocco to Egypt. North of the Mediterranean it breeds south of the valley of the Danube and east of the Adriatic, and in Southern Russia. Eastwards in Asia it breeds throughout Persia, Turkestan, and Southern Siberia as far north as the Baikal area and the valley of the Amoor. It is said also to breed in Japan.

BREEDING HABITS: In Europe the Ruddy Sheldrake is sedentary, but in Asia, where the climate of its breeding area is more rigorous, it leaves its summer haunts to winter in China, Burma, and India. It reaches its northern breeding grounds again in April or early May, migrating in flocks, but separating into scattered pairs for the nesting season. The Ruddy Sheldrake pairs for life. The favourite haunts of this species are reed-fringed rivers in which sandy islands occur, and which flow over wide fertile plains, also lagoons, and less frequently bare mountainous districts often far from water. The nest is made in a great variety of places, but almost invariably in a covered site. Sometimes holes in cliffs are selected, or burrows and clefts in the ground, even in the centre of a cornfield; whilst holes in trees and logs, and the deserted nests of birds of prey, are also chosen. Prjevalsky mentions fire-places in the houses of deserted Mongol villages as an occasional site ; whilst it has been found amongst a colony of Griffon Vultures, or near to nests of the Raven, Black Kite, and Egyptian Vulture in the cliffs. Usually the nest is not far from water, but instances are by no means rare in which it has been found long distances from it. The nest is made almost entirely of down, mixed with a few straws or bents.

RANGE OF EGG COLOURATION AND MEASUREMENT : The eggs of the Ruddy Sheldrake are from eight to sixteen in number, but eight or ten is the average clutch. They are smooth in grain, very fragile, and creamy-white. Average measurement, 2·7 inches in length by 1·8 inch in breadth. Incubation, said only to be performed by the female, but probably by both sexes, lasts thirty days.

DIAGNOSTIC CHARACTERS : The eggs of the Ruddy Sheldrake cannot be distinguished from those of the Common Sheldrake. Unfortunately the down of this species appears not to have been described, but will probably prove of some assistance in identifying the eggs.

Family ANATIDÆ.   Genus ANAS.
Sub-family ANATINÆ.

AMERICAN WIDGEON.

ANAS AMERICANA, *Gmelin*.

(British : Very rare abnormal autumn migrant.)

Single Brooded. Laying season, last half of May, in June, and possibly July.

BREEDING AREA : Northern Nearctic region. The American Wigeon breeds in Alaska, probably throughout British America, as far north as lat. 70°, and southwards to the most northern United States—Dakota, Minnesota, etc.

BREEDING HABITS: The American Wigeon reaches its more southerly breeding grounds in April or early in May, but it is about a month later in the high north. Although gregarious during the winter the flocks disperse in spring, and breed in scattered pairs. Of the pairing habits of this Duck nothing appears to be known. The favourite nesting haunts of the American Wigeon are swampy grounds, either on the treeless tundras, or near the prairie lakes, and rough marshy grounds studded with trees and bushes. The nest appears always to be made on the ground, often beneath the shelter of a tuft of rushes or coarse grass, or under a bush, a dry patch of ground being selected if possible. The nest is merely a hollow lined with dry grass or a few leaves, to which, however, is added a thick and abundant lining of down and a few feathers. The bird is a close sitter, remaining on the nest until almost trodden upon, and when flushed flying straight away, with little or no demonstration.

RANGE OF EGG COLOURATION AND MEASUREMENT: The eggs of the American Wigeon are from six to twelve in number. They are creamy-white or pale buff in colour. Average measurement, 2·2 inches in length by 1·5 inch in breadth. Incubation is performed by the female, but the duration of the period has not been determined.

DIAGNOSTIC CHARACTERS: The eggs of the American Wigeon cannot be distinguished from those of the Common Wigeon, and as the down is not described, it is impossible to say whether it is of any service in identifying the eggs. The locality is of value, for the Common Wigeon does not breed in the New World.

Family ANATIDÆ. Genus ANAS.
Sub-family *ANATINÆ*.

AMERICAN TEAL.

ANAS CAROLINENSIS, *Gmelin*.

(British : Very rare abnormal autumn migrant.)

Single Brooded probably. Laying season, June.

BREEDING AREA : Northern Nearctic region. The American Teal breeds in the Arctic and sub-arctic regions of North America, from the Aleutian Islands and Alaska eastwards to Greenland, and southwards through Canada to the more northern United States.

BREEDING HABITS : To its more northern breeding places the American Teal is only a summer visitor, returning to them in April and May. In winter it is gregarious, but during the breeding season lives in scattered pairs. Its favourite haunts are small pools, and swamps in which the open sheets of water are separated by strips of rough wet land, studded with tussocks of grass and rushes. It is probable that this Duck pairs for life, but nothing positive is known. The nest is invariably placed on the ground, usually amongst long coarse grass or rushes, often sheltered by a tuft of vegetation, or amongst willow thickets. It is merely a hollow carefully lined with scraps of dead herbage and quantities of down and feathers. The bird sits closely, when flushed flying straight away without demonstration.

RANGE OF EGG COLOURATION AND MEASUREMENT : The eggs of the American Teal are from eight to ten or even twelve in number, nine being an average clutch. They are smooth in grain, polished, and creamy-white. Average measurement, 1·8 inch in length by 1·3 inch in breadth. Incubation, performed by the female, lasts about three weeks.

DIAGNOSTIC CHARACTERS: The eggs of this Teal cannot be distinguished from those of the European Teal, or from those of the Blue-winged Teal. As the European Teal does not breed in the New World there is no chance of confusion between the eggs *in situ*, and it is probable that the down of the two American Teals is different, serving to identify the eggs, but as usual our transatlantic naturalists have failed to describe it. They are ready enough to tell us that the nests of these Ducks are warmly lined with down, but the characteristics of that down are beneath their consideration or description.

Family ANATIDÆ.
Sub-family *ANATINÆ*.

Genus ANAS.

BLUE-WINGED TEAL.

ANAS DISCORS, *Linnæus*.

(British: Very rare abnormal autumn migrant, of doubtful occurrence.)

Single Brooded probably. Laying season, May and early June.

BREEDING AREA: Central and Southern Nearctic region. The Blue-winged Teal breeds throughout the North-American continent up to about lat. 60°, and as far south as Florida and Northern Mexico (to the Tropic of Cancer). It is much rarer and more local west of the Rocky Mountains.

BREEDING HABITS: The Blue-winged Teal is not nearly so boreal a species as the American Teal, but frequents very similar localities. In northern areas it is a migrant, leaving for the south in autumn and returning in April and May. Its favourite breeding haunts are swampy ponds fringed with willows and coarse grass and rushes : ponds containing small islands are preferred. The nest is invariably made upon the ground, either amongst the dense vegetation on the bank of the pool, or on an island, beneath a tuft of grass or other herbage. The nest is a mere hollow, lined with scraps of dead vegetation and a copious supply of down and feathers. The bird sits closely, waiting till almost trodden upon before starting rapidly from the nest, and usually taking refuge some distance away amongst the vegetation, or settling on the open water.

RANGE OF EGG COLOURATION AND MEASUREMENT : The eggs of the Blue-winged Teal are from eight to twelve in number, ten being an average clutch. They are creamy-white, smooth and somewhat polished. Average measurement, 1·9 inch in length by 1·3 inch in breadth. Incubation, performed by the female, lasts about three weeks.

DIAGNOSTIC CHARACTERS : As the down of this species still remains undescribed, I am unable to give any character by which the eggs can be distinguished from those of allied birds.

Family ANATIDÆ. Genus FULIGULA.
Sub-family *FULIGULINÆ.*

RED-CRESTED POCHARD.

FULIGULA RUFINA (*Pallas*).

(British : Rare abnormal winter migrant.)

Single Brooded. Laying season, April to June, according to locality.

BREEDING AREA: South-western Palæarctic region. The Red-crested Pochard breeds locally in the Spanish Peninsula, principally in the eastern portion, in the Balearic Islands, Sardinia, Sicily, Italy, Central and Southern Germany, the valley of the Danube, and Southern Russia. South of the Mediterranean it breeds on the lakes of Northern Africa, principally in the north-west. Whether it breeds in Asia Minor or Armenia appears not to be known, but it certainly does so in Northern Persia and Turkestan.

BREEDING HABITS: In the more northerly portions of its distribution the Red-crested Pochard is a migrant, reaching its breeding haunts in April. This Duck is not only gregarious during winter, but is more or less social during summer, many pairs often nesting in close proximity. The bird probably pairs for life. Ponds, broads, and lakes with plenty of cover—rushes, willow and alder scrub, long grasses—round the margin, and especially such as contain islands, are the favourite nesting grounds of this Pochard. The nest is generally made amongst the flags, rushes, and coarse long grass near the water. Externally it is composed of dead rushes, leaves, bits of reed, and other vegetable refuse, and warmly lined with down and a few feathers. In some cases the old nest of a Coot or Moorhen is annexed. The bird is a close sitter, and when leaving the nest voluntarily covers the eggs with down, a proceeding common to most if not all species in this order.

RANGE OF EGG COLOURATION AND MEASUREMENT: The eggs of the Red-crested Pochard are from seven to ten in number. Specimens taken by Mr. Salvin in Algeria are described as of "a most brilliant fresh green colour when unblown; the contents were no sooner expelled, and the egg dry, than the delicate tints were gone, and their beauty sadly diminished." After being in collections for any length of time they may be described as grayish-olive. Average measurement, 2·3 inches in length by 1·6 inch in breadth. Incubation is performed by the female, but the duration of the period is unknown.

DIAGNOSTIC CHARACTERS: Unfortunately the down of this Duck does not appear to have been described, and we are left without a valuable clue to the correct identification of its eggs. They very closely resemble those of the Pochard, but always appear to be greener, approaching in colour those of the Golden Eye, but the down of the latter species is pale.

Family ANATIDÆ.
Sub-family *FULIGULINÆ.*

Genus FULIGULA.

WHITE-EYED POCHARD.

FULIGULA NYROCA (*Güldenstädt*).

(British : Rare abnormal spring, autumn, and winter migrant.)

Single Brooded. Laying season, end of April to June, according to latitude.

BREEDING AREA: Southern Palæarctic region and extreme North-western Oriental region. The White-eyed Pochard has a very extensive distribution, breeding throughout Europe from the Mediterranean northwards to Holland, the southern shores of the Baltic, Moscow, Kazan, and Ekaterinburg. South of the Mediterranean it breeds in North Africa from Morocco to Tunis, whilst eastwards it does so in Asia Minor, Armenia, Northern Persia, Turkestan, Cashmere, and probably through Mongolia, Manchooria, and the Amoor districts. It is not known to breed anywhere in Siberia, although it was seen by Finsch in the valley of the Obb as far north as the Arctic circle.

BREEDING HABITS : In the warmer portions of its distribution the White-eyed Pochard is sedentary, but in colder areas it is a bird of regular passage, reaching its breeding grounds in March and April, or early in May. The breeding season of this Duck varies a good deal according to locality. In Germany and the valley of the Danube it lays in May ; in Spain towards the end of April ; in Algeria not before June, which is also the date for Cashmere. The principal breeding haunts of this Pochard are slow-running rivers, ponds, broads, and lakes, especially such as contain islands, and where the banks are swampy and covered with a luxuriant growth of rushes, tall grass, shrubs, and other aquatic vegetation. The nest is generally made among the reeds and rushes on the banks of the pool, either on the land or on floating masses of rotten fallen vegetation or drifting weed. Sometimes it is made on a tuft or hassock of sedge or rush ; whilst Taczanowski records it being carefully concealed in a bush several feet from the ground. It is made of dry rushes, sedge, and other vegetable refuse, the finer materials being used for the interior, which is again lined with down and a few feathers plucked from the body of the female. Of the pairing habits of this Duck nothing appears to be known. It is a close sitter, but flies right away when flushed, covering its eggs, however, with down when leaving them voluntarily:

RANGE OF EGG COLOURATION AND MEASUREMENT: The eggs of the White-eyed Pochard are from eight to fourteen in number, ten being an average clutch. They are pale creamy-buff, sometimes with a perceptible tinge or suffusion of green. Average measurement, 2·1 inches in length by 1·49 inch in breadth. Incubation, performed by the female, lasts, according to Favier, thirty days; Naumann says twenty-two to twenty-three days.

DIAGNOSTIC CHARACTERS: The eggs of this Pochard very closely resemble those of the Gadwall and the Wigeon, but the down tufts, which are small and very dark brown without any pale tips, will serve to distinguish them. The eggs of the Harlequin Duck and the Smew also resemble them, but the breeding grounds do not impinge, the nests are different, and the shell weight of the Pochards is heavier than any of the rest, size for size.

Family ANATIDÆ. Genus FULIGULA.
Sub-family *FULIGULINÆ*.

SCAUP.

FULIGULA MARILA (*Linnæus*).

(British: Common autumn and winter migrant.)

Single Brooded. Laying season, May and June.

BREEDING AREA: Northern Nearctic and Palæarctic regions. The Scaup breeds sparingly in the Faroes, abundantly in Iceland, and throughout the Arctic regions of Europe and Asia, from the Atlantic to the Pacific, as far north as lat. 70°, and at high elevations on the mountains of Southern Scandinavia. In the New World it breeds as far north as lat. 70° from east to west, and as far south as the Hudson Bay Territory. Whether it breeds in Greenland is not yet determined.

BREEDING HABITS: The Scaup is a migrant, reaching its Arctic breeding grounds with the break up of the ice and the melting of the snow towards the end of May or early in June. During winter this Duck is gregarious enough, and even in summer continues sociable, many pairs nesting within a small area and collecting at certain spots to feed. This bird apparently pairs for life. Its favourite breeding grounds are the tundras near the lakes, which are fringed with rushes and grass, and often surrounded with thickets of birches, junipers, and willows. The nest is made upon the ground by the water-side among willows and junipers, or on a bank clothed with species of Ericaceæ, and studded with tufts of sedge and coarse grass. In Iceland it was found by Proctor amongst large stones near the water's edge. It is merely a hollow lined with dry grass, bits of sedge, and withered leaves, but warmly finished off with down, which increases in amount as the full clutch of eggs is laid. The bird is a close sitter, waiting until the last moment before quitting the eggs, which, however, are carefully covered for concealment when left voluntarily.

RANGE OF EGG COLOURATION AND MEASUREMENT: The eggs of the Scaup are from six to nine in number, but where several females lay in the same nest, as is sometimes the case, Dr. Kruper found as many as twenty-two. They are pale greenish-gray, and smooth in texture. Average measurement, 2·6 inches in length by 1·7 inch in breadth. Incubation is performed by the female, but the duration of the period is unknown.

DIAGNOSTIC CHARACTERS: The eggs of the Scaup

and those of the Pochard overlap in size and are the same in colour, but the down of the latter is grayer, that of the present species being large tufted and dark brown, with pale centres.

Family ANATIDÆ.
Sub-family *FULIGULINÆ.*

Genus FULIGULA.

HARLEQUIN DUCK.

FULIGULA HISTRIONICA (*Linnæus*).

(British : Very rare nomadic autumn migrant.)

Single Brooded.   Laying season, May and June.

BREEDING AREA: Eastern Palæarctic and Nearctic regions. The Harlequin Duck breeds in Iceland, Greenland south of the Arctic circle, and thence across the North American continent[1] from about the latitude of the Arctic circle south to lat. 45°. Westwards it breeds in the Aleutian Islands, and probably in Kamtschatka, the Stanavoi Mountains, the valley of the Amoor, and the Baikal area. There is no absolutely reliable evidence of this bird breeding anywhere in Western Siberia or on continental Europe.

BREEDING HABITS: The Harlequin Duck is a nomadic migrant, rarely straying much south of open water during winter. It is not gregarious during the breeding season, living in scattered pairs: the flocks of this Duck noted by Elliott in the Pribylov Islands being probably composed of immature non-breeding birds.

[1] It is rather remarkable that this Duck was never met with by Macfarlane during his long sojourn in the Arctic regions of America.

The favourite haunts of the Harlequin Duck are swift-flowing rivers and streams, such as are broken up into falls and rapids being preferred. It has been stated that this species nests in holes on the banks of rivers and in tree-trunks, but this is probably erroneous. Reliable authorities state that the nest is made upon the ground close to the edge of the stream, but I cannot find that it has ever been seen *in situ*, or described by a competent naturalist. Of the pairing habits of this beautiful Duck nothing apparently is known. It seems marvellous that the nesting economy of a bird breeding as near to our shores as Iceland should be so utterly unknown, or that what little information we do possess should be so unreliable.

RANGE OF EGG COLOURATION AND MEASUREMENT: The eggs of the Harlequin Duck are from eight to ten in number, six or seven being the average clutch. They are smooth in texture, with some polish, and according to Mr. Raine, who has received upwards of 200 eggs of this bird from his collector in Iceland within the past five years, are "deep rich buff, some having a yellowish tinge, others are pale buffy cinnamon." Average measurement, 2·2 inches in length by 1·7 inch in breadth. Incubation is apparently performed by the female, but the duration of the period is unknown.

DIAGNOSTIC CHARACTERS: The eggs of the Harlequin Duck somewhat closely resemble those of the Smew in colour, but may always be distinguished by their richer tint, larger size, and heavier shell-weight. The down is described by Mr. Raine as "dark grayish-brown," which is also a further point of distinction, that of the Smew being very pale grayish-white.

Family ANATIDÆ. Genus FULIGULA.
Sub-family FULIGULINÆ.

LONG-TAILED DUCK.

FULIGULA GLACIALIS (*Linnæus*).

(British : Fairly common autumn and winter migrant.)

Single Brooded. Laying season, from end of May to the first half of July, according to locality.

BREEDING AREA: Northern Nearctic and Palæarctic regions. The Long-tailed Duck breeds on the tundras and barren grounds above the limits of forest growth in Europe, Asia, and North America, as far north as land exists, and in a similar climate at high elevations in Scandinavia, Iceland, and perhaps the Faroes.

BREEDING HABITS: The Long-tailed Duck is a migrant, and returns to its breeding grounds in the Arctic regions as soon as open water can be found, when the ice on the great rivers breaks up, and the snow melts from the tundras. The favourite summer haunts of this Duck are the Arctic tundras and barren grounds of both hemispheres, which extend from the limits of forest growth to the Polar seas. Here it frequents the pools and lakes, especially such as are studded with islands. During winter the bird is gregarious, and even in the breeding season a certain amount of sociability is observable. Scattered pairs frequent the smaller pools, but the larger sheets of water are the resort of perhaps a dozen or twenty pairs. This Duck probably mates for life, but precise information is wanting. The nest is generally placed upon the ground in some sheltered nook, often amongst willow and birch scrub, or on the drifted rubbish left by the floods when the big northern rivers break up in spring, or amongst long grass. An island is selected if one is to be had. The nest is merely a hollow amongst the herbage, lined with an abundance of down and a few feathers plucked from the body of the female. The bird is a close sitter, and, when voluntarily leaving her eggs, covers them for concealment. The male assists the female in bringing up the brood, which is a very exceptional circumstance.

RANGE OF EGG COLOURATION AND MEASUREMENT: The eggs of the Long-tailed Duck are from seven to twelve in number, eight or nine being the average clutch.[1] They are smooth in texture, have some polish, and vary in colour from pale buffish-green to greenish-buff. Average measurement, 2·1 inches in length by 1·5 inch in breadth. Incubation is performed apparently by the female, but the duration of the period is unknown.

DIAGNOSTIC CHARACTERS: The eggs of the Long-tailed Duck cannot always be distinguished from those of the Pintail Duck, or even from those of the Mallard, but the down from the nest is a tolerably safe guide to their identification. The tufts are small, warm brown in colour, without any white tips.

[1] Macfarlane writes: "Considerably over one hundred nests were taken, and the eggs varied from five to seven, the latter being the maximum number recorded in any one instance," so that the clutches in the Nearctic region are probably smaller than those in the Palæarctic region. (Conf. *Proc. U. S. Nat. Mus.*, XIV., p. 421.)

Family ANATIDÆ. Genus FULIGULA.
Sub-family *FULIGULINÆ.*

VELVET SCOTER.

FULIGULA FUSCA (*Linnæus*).

(British : Fairly common autumn and winter migrant.)

Single Brooded. Laying season, end of June and early July.

BREEDING AREA : Northern Palæarctic region. The Velvet Scoter breeds in the Arctic and sub-arctic regions of Europe and Asia, as far north as lat. 69°, and as far south as the Baltic Provinces in the west and lat. 55° in the east. The Nearctic representative is *F. fusca velvetina.*

BREEDING HABITS: The Velvet Scoter is a migratory bird, coming south in winter, but retiring north again in spring to breed, reaching its nesting grounds with the break-up of the ice and the melting of the snow. Its breeding grounds are situated at some distance from the coast, as a rule, and not unfrequently the nest is made a long way from any water at all. It is gregarious during winter, but appears to live in scattered pairs only during the breeding season. Its favourite haunts are the tundras, usually in the vicinity of the great rivers and lakes. The nest is made upon the ground amongst scrub or the coarse vegetation of the tundra, or under the shelter of a rush-tuft or solitary bush near the water, but sometimes on a dry part of the moor. It is merely a hollow, in which a little dry grass, a few dead leaves or other vegetable fragments have been placed, and lined with down and a few feathers from the body of the female. The bird sits closely, and always covers the eggs for concealment when leaving them voluntarily.

RANGE OF EGG COLOURATION AND MEASUREMENT :

The eggs of the Velvet Scoter are from five to eight or nine in number, generally eight. They are smooth in texture, with little polish, and pale grayish-buff in colour. Average measurement, 2·8 inches in length by 1·9 inch in breadth. Incubation is performed by the female, but the duration of the period is unknown.

DIAGNOSTIC CHARACTERS: The size combined with the colour is a pretty safe guide to the determination of the eggs of this Scoter. They are larger than those of the Common Scoter and the Surf Scoter. The down tufts are also larger than the down tufts of the Common Scoter: brown shot with gray, and with indistinct pale centres.

Family ANATIDÆ. Genus FULIGULA.
Sub-family *FULIGULINÆ*.

SURF SCOTER.

FULIGULA PERSPICILLATA (*Linnæus*).

(British : Rare nomadic autumn and winter migrant.)

Single Brooded. Laying season, end of June and early July.

BREEDING AREA: Northern Nearctic region. The Surf Scoter breeds in the Arctic and sub-arctic regions of North America from the Atlantic to the Pacific, as far north as lat. 70°, and as far south as lat. 50°.

BREEDING HABITS: Like most other Arctic Ducks, the Surf Scoter appears at its breeding grounds in spring with the break-up of the ice, and as summer bursts with startling suddenness over the lonely forests and barrens. It is a very gregarious bird during winter, and even in summer, as soon as the females have scattered themselves over the suitable nesting grounds, the males begin again to flock and to keep company until the following spring. Of the pairing habits of this Duck nothing appears to be known. Its favourite breeding grounds are the open wooded areas studded with lakes and streams and swamps, as well as the bare treeless barren grounds, with their rush and grass-fringed pools and bogs. The nest is usually made near the water amongst scrub and coarse vegetation. MacFarlane and others have found the nest of this Scoter at the foot of pine trees concealed by the drooping lower branches; Audubon shot the female from a nest in a tussock of grass in a marsh. The nest is merely a hollow in the ground, lined with any scraps of vegetation that may chance to be near, and also with quantities of down and a few feathers plucked from the body of the female. The bird sits closely, and when leaving the nest to go and feed carefully covers the eggs, to shield them from the prying glances of enemies.

RANGE OF EGG COLOURATION AND MEASUREMENT: The eggs of the Surf Scoter are from five to eight in number, the latter, however, being exceptional. They are smooth in texture, with little polish, and pale grayish-buff in colour. Average measurement, 2·3 inches in length by 1·65 inch in breadth. Incubation is performed by the female, but the duration of the period is unknown.

DIAGNOSTIC CHARACTERS: The eggs of this Scoter are distinguished by their size and colour, being smaller than those of the Common Scoter and those of both forms of the Velvet Scoter. Whether the down is of any service as an additional means of identification I cannot say, as no naturalist appears ever to have described it.

Family ANATIDÆ. Genus CLANGULA.
Sub-family FULIGULINÆ.

BUFFEL-HEADED DUCK.

CLANGULA ALBEOLA (*Linnæus*).

(British : Very rare abnormal winter migrant.)

Single Brooded. Laying season, May and June.

BREEDING AREA: Northern Nearctic region. The Buffel-headed Duck breeds in the Arctic regions of North America up to the limits of forest growth, and as far south as the northern states of Maine and Wisconsin.

BREEDING HABITS: The Buffel-headed Duck is a migratory species, and returns to its northern breeding grounds as soon as the ice breaks up and it can obtain food. The favourite breeding haunts of this pretty Duck are wooded areas in which plenty of lakes and pools occur. Although gregarious during the winter it appears to live in scattered pairs during the nesting season, and the male seems to keep closer companionship with the female than is usual in this class of birds. The nest is made in a hollow tree, from fifteen to twenty feet above the ground, either a hole in a branch or in the trunk being selected. It would appear, however, that in districts where suitable trees are scarce a hole in a bank will be used. Mr. Raine (*op. cit.*, p. 62) gives a very interesting account of his finding a nest of this Duck containing twelve eggs in a gopher hole (the gopher is a little burrowing animal), made in the bank of a lake overgrown with bushes and honeycombed with the burrows of this quadruped. The duck flew from the hole and alighted on the adjoining lake, where it was joined by the drake. The eggs were laid on a bed of down three feet from the entrance to the burrow. Beyond lining the selected hole copiously with down this Duck makes no nest.

RANGE OF EGG COLOURATION AND MEASUREMENT: The eggs of the Buffel-headed Duck are from six to ten or twelve in number. They are smooth in texture, with some polish, and greenish-gray in colour. Mr. Raine describes his eggs as "warm dull buff colour, with a grayish tinge." Average measurement, 2·0 inches in length, by 1·45 inch in breadth. Incubation may possibly be performed by both sexes, but the matter still requires determination. The duration of the period is unknown.

DIAGNOSTIC CHARACTERS: The eggs of the Buffel-headed Duck somewhat closely resemble those of the Gadwall, but the situation of the nest is different, as is also, most probably, the down, although American naturalists have as yet omitted to describe it. Their colour readily distinguishes them from those of the Golden-eye and Barrow's Golden-eye; they are also smaller.

Family ANATIDÆ. Genus CLANGULA.
Sub-family *FULIGULINÆ*.

GOLDEN-EYE.

CLANGULA GLAUCION (*Linnæus*).

(British: Common autumn and winter migrant.)

Single Brooded. Laying season, May and June.

BREEDING AREA: Northern Nearctic and Palæarctic regions. The Golden-eye breeds in Iceland, and throughout the Arctic and sub-Arctic regions of Europe and Asia up to the limits of forest growth. In Europe it breeds as far south as Holstein, Brandenburg, Pomerania, and Prussia, and in Russia as far south as the Caucasus. In Asia it breeds throughout Siberia south of the limits of forest growth. In the New World (if we admit that the birds are identical) it breeds throughout British North America, and Alaska, south of the limits of forest growth.

BREEDING HABITS: The Golden-eye is a somewhat hardy species, and its migrations are consequently limited. It returns to its old breeding places as soon as they are free from ice, the date varying according to locality and state of the season. During winter the Golden-eye is more or less gregarious, and in many cases continues social throughout the breeding season, numbers of nests being made within a small area if suitable sites are to be had. This Duck probably pairs for life and returns to the same nesting place every season. The Golden-eye may frequently be seen to perch in a tree. The favourite breeding haunt of this Duck is open forest country where the trees are large and many of them decayed, and where the ground is broken up into swamps and lakes with the timber more or less thickly interspersed between them. The nest is usually made in a hollow tree as much as thirty feet from the ground, either in a hole in the trunk or in a hollow branch, the deserted hole of a Black Woodpecker sometimes being used. Naumann asserts that the nest is frequently made amongst rushes and other aquatic vegetation, or on the top of a pollard either near the water or at some distance from it: this is probably in districts where no hollow trees can be found. The Lapp and Finnish peasants place boxes and hollow logs for this Duck to breed in, regularly but judiciously removing the eggs. The partiality of this species for a nesting site near a waterfall or quick-flowing stream has been remarked by several observers. No nest beyond a thick warm bed of down is provided. The bird is a close sitter, and when flushed flies straight away at once.

RANGE OF EGG COLOURATION AND MEASUREMENT: The eggs of the Golden-eye are usually from ten to thirteen in number, but exceptionally as many as nineteen have been found, probably the produce of two females. They are smooth in texture, somewhat glossy, and bright grayish-green in colour. Average measurement, 2·3 inches in length, by 1·6 inch in breadth. The duration of the period of incubation is unknown. Whether both sexes incubate is still undetermined.

DIAGNOSTIC CHARACTERS: The size, green colour of the eggs, and situation of the nest combined with the colour of the down are sufficient to identify the eggs of the Golden-eye. The down is pale lavender-gray, with obscure pale centres. The eggs of Barrow's Golden-Eye are larger (average 2·4 by 1·72) and paler.

Family ANATIDÆ.　　　　　　　　　　Genus SOMATERIA.
Sub-family *FULIGULINÆ*.

STELLER'S EIDER.

SOMATERIA STELLERI (*Pallas*).

(British: Very rare nomadic autumn and winter migrant.)

Single Brooded. Laying season, end of June and early July.

BREEDING AREA: North-eastern Palæarctic region, and possibly the extreme North-western Palæarctic region. Steller's Eider is only known to breed normally and with certainty in Kamtschatka, on the islands in Bering Strait, and the Aleutian Islands; in the delta of the Lena, and on the Taimyr peninsula. It has been said to breed on the coast of Russian Finmark, and in the Varanger Fjord, but, if true, the occurrence is most exceptional.

BREEDING HABITS : Steller's Eider wanders no further from its breeding grounds than the severity of the weather compels. It is a nomadic migrant, only coming south to where open water can be found, and retiring north again with the break-up of the ice. It is gregarious during winter, and more or less social in summer, numbers of ducks nesting in close proximity, and the drakes swimming and feeding in company. The breeding haunts of this beautiful Eider are the rocky coasts of the Arctic Ocean. Of its pairing habits we know nothing. But little has been recorded of the nidification of this species ; indeed the only information we possess is that obtained by Von Middendorff in Siberia. He met with this Duck breeding in some numbers on the Taimyr peninsula, which is remarkable for being the most northerly continental land in the world. The nests were made on the tundra, and are described as deep hollows in the moss-clothed ground, lined with quantities of down plucked from the body of the female. The ducks sit very closely, the drakes swimming in attendance on the adjoining sea, and meeting their mates when they leave their nests for a brief period to feed.

RANGE OF EGG COLOURATION AND MEASUREMENT : The eggs of Steller's Eider are from seven to nine in number. They are smooth in texture, slightly polished, and pale buffish-green in colour. Average measurement, 2·35 inches in length, by 1·55 inch in breadth. Incubation is performed by the female, but the duration of the period is unknown.

DIAGNOSTIC CHARACTERS : The eggs of this bird are very rare in collections, and unfortunately they possess no character by which they can be always distinguished from those of the Pintail Duck. The down of this Eider is undescribed, but will doubtless prove sufficiently characteristic to render the identification of the eggs complete. Eggs of this species, unaccompanied by down, unless thoroughly well authenticated, are worthless as scientific specimens.

Family ANATIDÆ.   Genus SOMATERIA.
Sub-family FULIGULINÆ.

KING EIDER.

SOMATERIA SPECTABILIS (*Linnæus*).

(British : Probably breeds : Rare nomadic autumn and winter migrant.)

Single Brooded. Laying season, first half of July.
(June, if in British Islands.)

BREEDING AREA: Northern Nearctic and Palæarctic regions. The King Eider breeds on the islands off the coast of Northern Siberia, on Nova Zembla, Franz-Josef Land, probably Spitzbergen, and in Greenland, and on the islands and coasts of Arctic North America, perhaps as far north as land extends.

BREEDING HABITS: The King Eider is another nomadic species which only wanders south in winter as far as or little beyond where the ice compels it. It is more or less gregarious throughout the year, the females nesting in close company, and the males consorting together on the sea close to the breeding places. Of the pairing habits of this Eider nothing appears to be known, but it is probable that the birds mate for life. The breeding haunts of this species are the coasts of the Arctic Ocean, and various low rocky islands that stud that sea probably as far as the Pole. But little is known of its nesting habits. Von Middendorff met with this Duck breeding on the Taimyr peninsula; Captain Feilden observed it doing so on Floeberg Beach, in lat. $82\frac{1}{2}°$; whilst MacFarlane obtained nests on the Arctic coast of America near Franklin Bay, and states that occasionally a few pairs may be found breeding in close proximity to the Pacific Eider, *Somateria v. nigrum*. The nest, which is placed in a similar situation to that of the Common Eider, is merely a hollow, warmly lined with down from the body of the female, gradually accumulated as the eggs are laid.

RANGE OF EGG COLOURATION AND MEASUREMENT: The eggs of the King Eider, so far as is at present ascertained, are from four to six in number. They are smooth in texture, with little polish, and pale greenish-gray in colour. Average measurement, 2·6 inches in length, by 1·75 inch in breadth. Incubation is performed by the female, but the duration of the period is not known.

DIAGNOSTIC CHARACTERS: The eggs of this Eider are best distinguished by their size, being much smaller than those of the Common Eider, and larger than those of Steller's Eider. They somewhat closely resemble those of the Red-breasted Merganser, but are always greener. The down resembles that of the Common Eider, which varies from brownish-gray to grayish-brown, with obscure pale centres.

Family ANATIDÆ. Genus MERGUS.
Sub-family MERGINÆ.

HOODED MERGANSER.

MERGUS CUCULLATUS, *Linnæus*.

(British : Rare nomadic autumn and winter migrant.)

Single Brooded. Laying season, May and June.

BREEDING AREA: Northern Nearctic region. The Hooded Merganser breeds in Arctic and north temperate North America, from the Atlantic to the Pacific, as far north as the Arctic Circle, and as far south as lat. 45°.

BREEDING HABITS: The Hooded Merganser returns to its summer haunts as soon as they are free from ice. In winter this species is gregarious, but during the nesting season the flocks disperse and live in scattered pairs. The favourite breeding grounds of this Merganser are wooded areas, in which the timber is old, where lakes, streams, and swamps occur amongst the trees and break the monotony of the forests. It probably mates for life, although definite information is wanting. It should be an easy matter to learn whether the same nesting places are resorted to each year. The nest is made in a hole in a tree, or in a hollow, fallen log. It is probable that the bird makes use of a hole in a bank in districts where the trees are unsuitable. No nest is made beyond a plentiful lining of down, which gradually accumulates as the eggs are laid. The bird is a close sitter, but when flushed flies straight off to the nearest water without any demonstration of anxiety.

RANGE OF EGG COLOURATION AND MEASUREMENT: The eggs of the Hooded Merganser are from five to nine in number. They are smooth in texture, polished, remarkably rotund—a shape, it may be remarked, very prevalent among birds nesting in holes where space is limited—and pure white when blown. Before the contents are removed they have a beautiful pearly appearance. Average measurement, 2·1 inches in length, by 1·7 inch in breadth. Incubation is performed, it is said, by the female alone, and lasts thirty-one days. Further research may show that the male bird occasionally sits upon the eggs.

DIAGNOSTIC CHARACTERS: As a rule the eggs of the Hooded Merganser are larger than those of the Wood Duck—a species that also nests in hollow trees—but they should always be carefully identified. The down is an important point, and is very pale gray in colour. The colour of the eggs will readily serve to distinguish them from those of the Smew, the Buffel-headed Duck, and the Golden-eye.

Family ANATIDÆ.
Sub-family *MERGINÆ*.

Genus MERGUS.

SMEW.

MERGUS ALBELLUS, *Linnæus*.

(British: Rare nomadic autumn and winter migrant.)

Single Brooded. Laying season, end of June and in July.

BREEDING AREA: Northern Palæarctic region. The Smew breeds in Finnish Lapland, and in Russia as far north as the Arctic Circle, and as far south as about lat. 60° in the west, and the valleys of the Kama and Lower Volga in the east. In Asia it breeds across Northern Siberia, south of the Arctic Circle or the limits of forest growth, but its southern limits appear to be undetermined.

BREEDING HABITS: The Smew is a nomadic migrant, a bird that wanders no further south of its breeding grounds than the snow and ice compel it, returning to them at the first opportunity, when the south wind banishes winter, towards the end of May or early in June. Its favourite breeding haunts are the forest districts in which the timber is of mature growth, studded with lakes and swamps, and intersected by streams. In winter it is more or less gregarious, but during the breeding season lives in scattered pairs, probably as much because suitable nesting sites do not admit of many birds incubating in company, as from any inclination to breed in solitude. Of the pairing habits of the Smew nothing appears to be known, but probably the bird mates for life, and returns regularly to one spot to breed. The nest is made either in a hole in a tree trunk or a branch, or in a hollow log or stump. No nest beyond a warm lining of down is prepared for the eggs, the first being deposited upon the wood dust at the bottom of the hole. The Smew is a close sitter, but when flushed flies straight to the nearest water without any demonstration.

RANGE OF EGG COLOURATION AND MEASUREMENT: The eggs of the Smew were first made known to science by Wolley. They were sent to him by his collector, who had obtained a nest with the female bird from a hole in a rotten birch trunk in Russian Lapland. Messrs. Seebohm and Harvie-Brown procured others in the valley of the Petchora, during the summer of 1875. They are from seven to eight or nine in number, smooth in texture, somewhat polished, and creamy-white in colour. Average measurement, 2·0 inches in length, by 1·47 inch in breadth. Incubation is apparently performed by the female, but the duration of the period is unknown. It is possible that the male may occasionally sit upon the eggs.

DIAGNOSTIC CHARACTERS: The eggs of the Smew cannot be distinguished from those of the Wigeon, so far as colour is concerned, but though they are about the same size they are proportionately much heavier in shell weight. The down, which is pale grayish-white, is a further aid to their correct identification.

Family PHŒNICOPTERIDÆ.   Genus PHŒNICOPTERUS.

FLAMINGO.

PHŒNICOPTERUS ROSEUS, *Pallas.*

(British : Very rare abnormal autumn migrant.)

Single Brooded. Laying season, end of May and early June.

BREEDING AREA: South-western Palæarctic region and northern Ethiopian region. The Flamingo breeds in Spain, Southern France (marismas of the Guadalquivir, marshes of the Camargue, etc.), and various other portions of Southern Europe, as far east as the low swampy northern shores of the Caspian. Eastwards in Asia it certainly breeds on the Kirghiz Steppes and in Persia, but how far north and east is very imperfectly known, some authorities stating that it occurs as far as the Baikal area, which seems probable, seeing that it is a winter visitor to India. South of the Mediterranean it breeds in Algeria and Tunis. It may possibly breed on the Canaries and Cape Verd Islands, although no evidence of the fact was obtained by Mr. Meade Waldo in the former group; but in the latter locality nests were actually seen by Captain Dampier (*Collect. Voyages*, i. pp. 70, 71; 1729): later information is wanting. The breeding range of this species in the Ethiopian region still remains undetermined, but the bird is known to nest in Central Africa.

BREEDING HABITS: In most if not all parts of its range the Flamingo appears to be migratory, but the dates of its annual movements have been little recorded. The favourite breeding haunts of this bird are vast mud flats and the low islands in deltas—the low flat shores of lakes and inland seas where the water is shallow. Of the pairing habits of the Flamingo nothing appears to have been observed. At all seasons it is a gregarious bird, and seems always to breed in colonies of varying size. By far the best description of the breeding habits of this interesting bird is that recorded in the *Ibis* by Mr. Abel Chapman, who visited a vast colony in the marismas of the Guadalquivir. The colony was situated on some low mud islands, and long before the nests were reached "the strange forms," says Mr. Chapman, "of hundreds of Flamingoes met one's eye in every direction—some in groups or in dense masses, others with rigidly outstretched neck and legs flying in short strings, or larger flights 'glinting' in the sunlight like a pink cloud. Many pairs of old red birds were observed to be accompanied by a single white (immature) one. But the most extraordinary effect was produced by the more distant herds, the immense numbers of which formed an almost unbroken white horizon, a sort of thin white line separating sea and sky round a great part of the circle." He further writes: "On reaching the spot we found a perfect mass of nests; the low mud plateau was crowded with them as thickly as the space permitted. These nests had little or no height: some were raised two or three inches, a few might be five or six inches; but the majority were merely circular bulwarks of mud, with the impression of the bird's legs distinctly marked on it. The general aspect of the plateau was not unlike a large table covered with plates. In the centre was a deep hole full of muddy water, which, from the gouged appearance of its sides, appeared to be used as a reservoir for nest-making materials. Scattered all round this main colony were numerous single nests rising out of the water and evidently built up from the bottom. Here and there two or three or more of these were joined together—'semi-detached,' so to speak; these separate nests rose some six or eight inches above the water-level, and were about fifteen inches across. The water was about twelve or fifteen inches deep." Although hundreds of Flamingoes were seen sitting on their nests, "their long red legs doubled under their bodies, the knees projecting as far as, or beyond, the tail, and their graceful necks neatly curled away among their back feathers, like a sitting Swan, with their heads resting on their breasts," none of them contained any eggs on the 11th of May, and these were not obtained until a fortnight later. It may here be remarked that for years it was believed that the Flamingo incubated its eggs whilst sitting astride of its conical nest. Waterton, I believe, was the first naturalist to expose this error; Mr. Chapman's observations confirm the exposure and set all doubt at rest. The birds are wary enough at the breeding places, posting sentries on the outskirts of the colony, which raise the alarm and warn the sitting birds of impending danger.

RANGE OF EGG COLOURATION AND MEASUREMENT: The eggs of the Flamingo are two in number; in very exceptional cases one. They are rough in texture, somewhat pyriform in shape, and pale greenish-blue without markings. Most of this colour, however, is concealed by a thick coating of chalky-white shell. Average measurement, 3·7 inches in length, by 2·3 inches in breadth. Whether the female alone performs the duty of incubation is undetermined, but seems probable. Incubation is said by Brehm to last from 30 to 32 days.

DIAGNOSTIC CHARACTERS: The eggs of the Flamingo cannot readily be confused with those of any other Palæarctic species, their size and shape, combined with their chalky appearance and green under surface, rendering confusion almost impossible.

Family IBIDIDÆ. Genus PLEGADIS.

GLOSSY IBIS.

PLEGADIS FALCINELLUS (*Linnæus*).

(British : Rare abnormal spring and autumn migrant.)

Single Brooded. Laying season, May and June.

BREEDING AREA: South-western Palæarctic region, and Oriental region. The Glossy Ibis breeds in Spain and in the delta of the Rhone. Eastwards it breeds in the valleys of the Danube—especially the Obedska *bara* —and the Volga as far north as lat. 48°, in the plains of the Caucasus, and throughout Turkestan and south-west Siberia, in suitable localities as far north as lat. 48°. It also breeds in the valley of the Indus and in Ceylon. South of the Mediterranean it is only known to breed with certainty in North-west Africa, and even there its distribution appears to be poorly defined.

BREEDING HABITS: The Glossy Ibis arrives at its breeding haunts in Europe towards the end of March or in April. Its favourite haunts are swamps and extensive marshes, rivers subject to periodical inundations, which every year are flooded, and turn the forests of willows and alders on their banks into haunts almost impenetrable to man. The most remarkable breeding place of this Ibis in Europe is situated on the Obedska *bara* in Sclavonia, a vast region of reeds and willows and swamps and inundated forests of alder and other trees, flooded every year by the rising of the Danube which flows through it. This Ibis is gregarious, and breeds in colonies of varying size, often in company with various species of Herons, Cormorants, Darters, and other swamp-loving birds. In Europe the nests are usually made in partly submerged willows; in India Mr. Doig found them in kundy trees; in Ceylon Col. Legge states that they were built in thorny trees growing in the half-dried bed of a small pool. The nests of several other species breeding in the colony may often be found on the same tree. The nests are flat and made of sticks and dry reeds, but those found by Col. Legge were mostly made of twigs and grass roots. In Europe where the selected trees are small— —little more than bushes—the nest is never far from the water, and often within a few inches of it, but in Ceylon where the trees are high the birds build at a much greater elevation. Of the pairing habits of this Ibis nothing is known.

RANGE OF EGG COLOURATION AND MEASUREMENT: The eggs of the Glossy Ibis are three or four in number, generally three. They are a clear greenish-blue—almost turquoise-blue—in colour, rough in grain, the shell being minutely pitted. Average measurement, 2·1 inches in length, by 1·5 inch in breadth. The duration of the incubation period is unknown, as is also which sex performs the task.

DIAGNOSTIC CHARACTERS: The eggs of the Glossy Ibis are readily distinguished from those of every other European bird by their size, blue colour, and pitted surface.

Family PLATALEIDÆ.   Genus PLATALEA.

SPOONBILL.

PLATALEA LEUCORODIA, *Linnæus*.

(British : Formerly bred : Occasional straggler on migration.)

Single brooded.   Laying season in Europe, May.

BREEDING AREA : Southern Palæarctic region. Northern Ethiopian region, and Oriental region. The Spoonbill still continues to breed in Holland and the South of Spain. Eastwards it becomes more plentiful in the valley of the Danube, and breeds in the delta of the Volga and in the Aral basin. Eastwards it breeds in Asia Minor, Turkestan, and West Siberia, where it was met with by Finsch as far north as lat. 48°. It also breeds regularly in Southern Dauria, the valley of the Amoor, and South-eastern Mongolia ; whilst southwards it does so over the whole of India and Ceylon. South of the Mediterranean it breeds in suitable localities throughout Africa as far south as the Soudan, and, it is said, the Dahalak Archipelago (lat. 16°) in the Red Sea.

BREEDING HABITS : The Spoonbill arrives at its more southerly breeding places in Europe about the middle of April, but is a fortnight or three weeks later in the north. It is a gregarious bird, breeding in colonies of varying size, keeping to itself and not intermixing with the Herons, Ibises, and Cormorants that may also be nesting in the same locality. There can be little doubt that the Spoonbill mates for life, for it returns yearly to the wonted place to breed, repairing or renewing the old nests as may be required. The usual summer haunts of this bird are swamps, lakes with shallow margins covered with reeds and other aquatic vegetation, and the partly submerged forests of willow and alder trees on the banks of rivers that are flooded every

O

spring. The nests are either made upon the ground—as is almost always the case at the Horster Meer in Holland—or upon low willow or alder bushes, or, as in India, on the summits of lofty trees. When in a tree the nest is much larger and more compact than when made amongst tussocks of grass upon the ground. In the latter situation the nest is made of a few sticks and quantities of dead reeds, and lined with dry grass; but when in bushes or trees it is a large pile of sticks, sometimes as much as a yard across and a foot in height. The cavity containing the eggs, in these latter nests, is shallow, and generally lined with dry grass. When disturbed at the nests the old birds rise and wheel silently about above the place, many of them flying right away without any further demonstration.

RANGE OF EGG COLOURATION AND MEASUREMENT: The eggs of the Spoonbill are four or five in number, generally the former. They are coarse and chalky in grain, without polish, and white in ground colour, somewhat sparingly spotted and blotted with reddish-brown, and with a few underlying markings of pale gray. They vary considerably in shape and colour, some being oval, others round, others pyriform. On some eggs the spots are small and streaky, and distributed here and there over the entire surface; on others the markings are congregated in a zone round the larger end; others have a few blurred blotches amongst the smaller spots and short streaks. Average measurement, 2·5 inches in length, by 1·8 inch in breadth. The duration of the incubation period is unknown, as is also the sex which performs the task.

DIAGNOSTIC CHARACTERS: The chalky grain, white ground colour, and brown markings readily distinguish the eggs of the Spoonbill from those of all other European species.

Family CICONIIDÆ.   Genus CICONIA.

WHITE STORK.

CICONIA ALBA, *Brisson*.

(British : Rare abnormal spring and autumn migrant.)
Single Brooded. Laying season, March to May according to locality.

BREEDING AREA : Western Palæarctic region. The White Stork breeds in Southern Sweden, in Denmark, Holland, Germany, and the Spanish peninsula. It also breeds in Austria, the Danubian provinces, Turkey, Greece, and throughout Russia as far north as the Baltic provinces ; in the Caucasus, Armenia, Asia Minor, Palestine, Persia, and Turkestan, as far east as Yarkand. South of the Mediterranean it breeds commonly in North-west Africa from Morocco to Tunis.

BREEDING HABITS : The White Stork throughout its vast breeding area is a migrant, arriving at its more southerly nesting places in March, but a month or more later at its northern ones. The arrival of the Stork in countries where the bird is common, is looked for as eagerly as the appearance of the Cuckoo and the Swallow in our own. It is a homely species, tame and confiding, because left unmolested, and breeding commonly in villages and towns, or near farmsteads. Indeed, in many localities a platform is erected for its accommodation by the farmer or the villager. In some places the White Stork is gregarious and breeds in colonies, as I remarked in Algeria, but in others it only occurs in scattered pairs. A village will frequently contain from six to a dozen nests, and in Algeria, in Batna, I counted several from the window of my hotel. The White Stork pairs for life, and returns each season to its old nest. This may either be built upon the roof of a house or other building, the tower of a mosque or a church, the ledge of a precipice (as I remarked at Constantine), or in the branches of a tree. The nest varies a good deal in size, the largest structures (five or six feet in height) being the oldest, and the accumulation of many years. It is generally a huge pile of sticks four or five feet across, amongst which twigs, dead reeds, and lumps of earth or clay are mixed; the cavity containing the eggs is shallow, and lined with almost any soft material that can be procured—dry grass, rags, feathers, straws, masses of hair, bits of paper, moss, or wool. The Stork sits closely and very tamely, with its long legs folded up beneath its body, and its equally long neck drawn in between the shoulders. If disturbed it wheels round above the spot, or glides to and fro past the face of the cliff, but quite silent.

RANGE OF EGG COLOURATION AND MEASUREMENT: The eggs of the White Stork are from three to five in number. They are rough in grain, unpolished, and pure white. Average measurement, 2·8 inches in length, by 2·1 inches in breadth. Incubation, performed chiefly, if not entirely, by the female, lasts from twenty-eight to thirty-one days.

DIAGNOSTIC CHARACTERS: The eggs of the White Stork are slightly larger than those of the Black Stork, and when empty and held up to the light are yellowish-white inside, those of the latter bird being green.

Family CICONIIDÆ.   Genus CICONIA.
BLACK STORK.
CICONIA NIGRA (*Linnæus*).

(British : Very rare abnormal spring and autumn migrant.)

Single Brooded.   Laying season, April.

BREEDING AREA: Southern Palæarctic region. The Black Stork breeds sparingly in Southern Sweden, much more commonly in Hanover, Pomerania, and Prussia (Herr E. Hartert states that it is nowhere as abundant in Germany as in East Prussia). It also breeds in Poland, the valley of the Danube, Spain, Central and Southern Russia, Turkey, and the Caucasus. Southwards it is said to breed in Palestine, and does so in Persia, Turkestan, and Southern Siberia, reaching as high as lat. 55° in the valley of the Obb; eastwards through the Baikal area, the Amoor valley, and Northern China. South of the Mediterranean I can find no definite evidence of its breeding anywhere on the African continent.

BREEDING HABITS: The Black Stork is a regular migrant, arriving in its more southerly summer haunts in March, but nearly a month later in the north. It is a much wilder and more wary bird than the White Stork, and never nests in towns or much-frequented places, at least in Europe. It is also solitary in its habits, each pair keeping to one particular locality, although an instance is on record where two pairs bred in company. The favourite breeding haunts of the Black Stork are large little-frequented forests, especially such as are swampy or near to marshes in which the bird can find food. This Stork also pairs for life, returning to the same nest year after year. The nest is generally made in a large and lofty tree, on a flat horizontal branch, or in a fork near the trunk, but sometimes it is built in a cleft of the rocks or on the cliffs. Occasionally it has been known in a tree on the old nest of a White-tailed Eagle. It is a ponderous structure from four to six feet across, very flat, and made chiefly of sticks of all sizes and thicknesses. The cavity containing the eggs is very shallow, and lined with tufts of green moss. This lining is renewed each year. The bird is a close sitter, especially when incubation is far advanced, and when flushed rises into the air and wheels round and round above the tree, anxiously awaiting the fate of the nest, usually being joined by its mate.

RANGE OF EGG COLOURATION AND MEASUREMENT: The eggs of the Black Stork are from three to five in number. They are rough in texture, porous, with little or no polish, and pure white. Average measurement, 2·6 inches in length, by 2·0 inches in breadth. Incubation, performed chiefly, if not entirely, by the female, lasts about a month.

DIAGNOSTIC CHARACTERS: The eggs of the Black Stork are smaller than those of the White Stork, and when empty and held up to the light are green inside, those of the latter bird being yellowish-white.

Family ARDEIDÆ.   Genus ARDETTA.

LITTLE BITTERN.

ARDETTA MINUTA (*Linnæus*).

(British : Possibly breeds : Spring and autumn coasting migrant.)

Single Brooded. Laying season, May and June.

BREEDING AREA : South-western Palæarctic region, north-eastern Ethiopian region, and north-western Oriental region. The Little Bittern breeds throughout Europe in suitable localities south of the Baltic. It also breeds in Asia Minor, Palestine, Persia, Baluchistan, North-west India (including Cashmere), and North-western Turkestan. South of the Mediterranean it breeds in the Azores, Madeira, and Northern Africa, from Morocco to Egypt, but principally in the north-west.

BREEDING HABITS : The Little Bittern arrives at its European breeding places in the south during March and April, but not until May in the north. It is a shy, skulking, solitary species, frequenting marshes and swamps and the reed and rush-fringed margins of pools. Of the pairing habits of this bird nothing appears to be known. The nest is built amongst the dense aquatic vegetation, sometimes amongst the belt of reeds at some distance from the shore, half floating in the stagnant shallow water, or on the bank in rushes and coarse grass. Less frequently it is made on the flat top of a pollard willow ; whilst the bird has even been known to make use of the old nest of a Magpie built in a tree near the swamps. In India the favourite situation is amongst wild rice or rushes. The nest is a large slovenly mass of half-rotten vegetation, the cup, which is shallow and saucer-like, being made of drier and finer material, such as the dead flowers and leaves of reeds, fine grass, etc. The nests made in India do not appear to be so elaborate. The bird is a close sitter, but usually slips off the nest unseen and hides itself in the dense cover, making no demonstration of anxiety for the eggs.

RANGE OF EGG COLOURATION AND MEASUREMENT: The eggs of the Little Bittern are usually four or five in number, but nine have been recorded ! They are oval in shape, chalky in texture, minutely pitted, and pure white. Average measurement, 1·4 inch in length, by 1·0 inch in breadth. Incubation, performed chiefly, if not entirely, by the female, lasts, on the authority of Naumann, sixteen or seventeen days.

DIAGNOSTIC CHARACTERS: The rough chalky and pitted shell, small size, and white colour readily distinguish the eggs of the Little Bittern from those of all species likely to be confused with them.

Family ARDEIDÆ.　　　　　　　　　　Genus BOTAURUS.

AMERICAN BITTERN.

BOTAURUS LENTIGINOSUS (*Montagu*).

(British : Very rare abnormal spring and autumn migrant.)

Single Brooded.　Laying season, May and June.

BREEDING AREA: Nearctic region except the extreme north. The American Bittern breeds throughout British North America south of about lat. 58°, and over the whole of the United States down to Texas.

BREEDING HABITS: In the northern portions of its distribution the American Bittern is a migrant, leaving in autumn and returning in April and early May. It is a shy, skulking, solitary species, much more often heard than seen, and keeping close to the dense cover of the haunts it frequents. Of the pairing habits of this Bittern nothing apparently is known. Its favourite nesting haunts are marshes and swamps and the dense belts of aquatic vegetation round the more open waters. Much difference of opinion prevails concerning the nesting habits of the American Bittern. Some writers assert that the bird breeds in colonies, others that it lives in solitary pairs. There can be no question, to my mind, that the latter view is the correct one. The nest is made in the swamps, either on the ground amongst the aquatic vegetation, or more frequently amongst the rushes in shallow water a little distance from the shore. It has been said that the nest is built in trees and bushes, and that the eggs are laid on the bare ground or under a bush, but all these statements may, I think, be safely discarded. The nest is a large bulky structure, more than a foot in height, and composed of half-rotten sedges and rushes, the saucer-like cavity at the top being lined with the driest and finest material. The bird is a close sitter, and when flushed, hurriedly leaves the nest and seeks seclusion in the nearest cover.

RANGE OF EGG COLOURATION AND MEASUREMENT: The eggs of the American Bittern are usually four or five in number, but it is said that seven are exceptionally found. They are uniform brownish-olive or buff. Average measurement, 2·0 inches in length, by 1·5 inch in breadth. The duration of the period of incubation is unknown, also which parent performs the duty.

DIAGNOSTIC CHARACTERS: The eggs of the American Bittern are not easily confused with those of any other species breeding in the Nearctic region, but cannot be distinguished from those of the Palæarctic or Common Bittern. The *locality* of the eggs is sufficient therefore to determine their species.

Family ARDEIDÆ.   Genus NYCTICORAX.

NIGHT HERON.

NYCTICORAX GRISEUS (*Linnæus*).

(British : Rare abnormal spring and autumn migrant.)

Single Brooded. Laying season in Europe, end of April and in May and June.

BREEDING AREA: Southern and western Palæarctic region, Nearctic region, Ethiopian region, Oriental region, and northern Neotropical region. The Night Heron may yet breed sparingly in Germany and Holland, but most of its old nesting places are deserted. It still however breeds in Southern France, the Spanish peninsula, Northern Italy, Sardinia, the valley of the Danube, South-eastern Austria, Southern Russia, and the Caucasus. It also breeds in Armenia, Asia Minor, Palestine, Persia, Western Turkestan, Mongolia, China, and the south island of Japan ; the Burma peninsula, Ceylon, and India. South of the Mediterranean it breeds throughout Africa in suitable districts down to the Cape Colony. In the New World it breeds over the United States and the extreme south of Canada, and as far south as Mexico, Central America, and Brazil.

BREEDING HABITS: The Night Heron is only a summer visitor to Europe, reaching its breeding grounds during April and May. The favourite breeding grounds of this Heron in Europe are swamps and the submerged willow and alder forests on the banks of rivers which are flooded every spring. This bird breeds abundantly in various parts of the valley of the Danube in company with other birds of the same family. It is gregarious during the breeding season, also social. In most parts of the world the Night Heron habitually nests in trees, but in America it is said occasionally to breed on the ground in marshes and amongst rice swamps. In Europe low willow and alder trees are occupied, but in India taller trees are selected, and Swinhoe mentions a colony of this Heron which was established in some old banyan trees in the courtyard of the great Honam Temple at Canton. Bushes are however used in China as well as in Europe, the same naturalist describing an enormous colony of this bird in such places near another temple, the nests being placed on every available branch, sometimes only a few feet from the ground. The nest is somewhat small, flat and shallow, and made of sticks. The nests observed by Mr. Seebohm in the valley of the Danube are described by him as having the sticks arranged like radii from the centre, not in arcs round it. When in swamps it would appear that reeds and stalks of other aquatic vegetation are used instead of sticks. When disturbed at their colony the birds become very anxious, fluttering off their nests, and flying to and fro in noisy alarm.

RANGE OF EGG COLOURATION AND MEASUREMENT : The eggs of the Night Heron are from three to five in number. They are chalky in texture, without polish, and bluish-green in colour. Average measurement, 2·0 inches in length, by 1·4 inch in breadth. Incubation is apparently performed by the female alone, but the duration of the period is unknown. According to

Swinhoe's experience of this species in China the eggs are laid at intervals, and sat upon at once.

DIAGNOSTIC CHARACTERS: The eggs of the Night Heron cannot always be distinguished from those of the Little Egret, but are generally much larger. Eggs of the Buff-backed Heron also resemble them in size, but are always much paler in colour. The eggs of all these smaller herons require the most careful identification.

Family ARDEIDÆ.   Genus ARDEA.

BUFF-BACKED HERON.

ARDEA BUBULCUS, *Audouin.*

(British: Very rare abnormal spring and autumn migrant.)

Single Brooded. Laying season, May and June.
(N. Hemisphere.)

BREEDING AREA: Ethiopian region and extreme south-western Palæarctic region (Spain and Palestine). The Buff-backed Heron breeds commonly in the southern portions of the Spanish peninsula (in the marshes of Andalucia), and throughout Africa, from Morocco to Egypt in the north, down to the Cape Colony and Madagascar in the south.

BREEDING HABITS: The Buff-backed Heron is a summer migrant to the only part of Europe in which it breeds, reaching Andalucia in March; elsewhere it appears to be sedentary. This Heron is gregarious like so many of its congeners, breeding in colonies of varying size. Of its pairing habits nothing apparently is known. Its breeding haunts are not only in swampy woods, but in woods where the ground is dry; in treeless districts the bird breeds amongst reeds, and in Egypt, according to Heuglin, sometimes even in gardens. A large colony visited by Mr. Gurney in Algeria was situated in a bed of dead tamarisks only a few feet above the water. The nest is flat, shallow, and made of sticks, when in trees or bushes, and composed of dead reeds and other aquatic refuse when in marshes and swamps where trees are absent. In its behaviour at the nest the Buff-backed Heron does not differ from its allies, rising in a slow, awkward manner, and hovering noiselessly above the colony until the alarm subsides.

RANGE OF EGG COLOURATION AND MEASUREMENT: The eggs of the Buff-backed Heron are from three to five in number. They are chalky in texture, oval in form, without polish, and almost white, with a scarcely perceptible tinge of bluish-green. Average measurement, 1·8 inch in length, by 1·3 inch in breadth. The duration of the period of incubation is unknown, as is also which sex performs the duty.

DIAGNOSTIC CHARACTERS: The size combined with the pale colour sufficiently distinguish the eggs of the Buff-backed Heron from those of other European species. Unfortunately, however, they cannot be distinguished from eggs of Montagu's Harrier. Except in the Spanish peninsula, however, the breeding areas of these two birds do not impinge.

Family ARDEIDÆ. Genus ARDEA.

SQUACCO HERON.

ARDEA COMATA, *Pallas*.

(British : Rare abnormal spring and summer migrant.)

Single Brooded. Laying season, latter half of May, and in June.

BREEDING AREA : South-western Palæarctic region, and Ethiopian region. The Squacco Heron breeds in the Spanish peninsula, throughout the valley of the Danube in suitable localities, Southern Russia, the basin of the Caspian, and Syria. South of the Mediterranean it breeds throughout Africa, from Morocco to Egypt in the north, down to Namaqua Land, Natal, and probably Madagascar.

BREEDING HABITS : The Squacco Heron is only a summer visitor to Europe, reaching its breeding grounds during April or early in May. Like its congeners it is gregarious and social, breeding in colonies composed of various species. Of its pairing habits nothing appears to be known. Its breeding haunts are the same as those of the Night Heron and the Little Egret, flooded willow and alder forests, and in treeless districts amongst the reeds in swamps and marshes. Mr. Seebohm gives a very graphic account of his visit to the vast colony of Herons in the valley of the Danube near Hirsova. This colony is situated in a flooded forest of pollard willows, and was estimated to contain some five thousand nests of various species, chiefly Herons. The nest of the Squacco Heron he found made on precisely the same model as that of the Night Heron—twigs arranged from a centre like radii. They were slight, but rather deep, and the eggs could be seen through them from below.

Nests of this species made in swamps are composed of dead reeds and other aquatic herbage, and are larger and more bulky than the nests made in trees. The actions of this species at the nest, and when the colony is invaded, are precisely the same as those of the preceding species.

RANGE OF EGG COLOURATION AND MEASUREMENT: The eggs of the Squacco Heron are from four to six in number. They are chalky in texture, unpolished, and greenish-blue. Average measurement, 1·5 inch in length, by 1·1 inch in breadth. The duration of the period of incubation is unknown, as is also the sex which performs the duty.

DIAGNOSTIC CHARACTERS: The eggs of the Squacco Heron cannot readily be confused with those of any other European species, their size and colour combined readily distinguishing them.

Family ARDEIDÆ. Genus ARDEA.

LITTLE EGRET.

ARDEA GARZETTA, *Linnæus*.

(British : Very rare abnormal spring and summer migrant.)

Single Brooded. Laying season, May and June.

BREEDING AREA: Southern Palæarctic region, Ethiopian and Oriental regions. The Little Egret breeds in Europe in the Spanish peninsula, in Southern France (Rhone delta), Sardinia, Sicily, the valley of the Danube, and Southern Russia. In Asia it breeds in Asia Minor, Palestine, Persia, India and Ceylon, Burma, China and Japan. In Africa it breeds in suitable districts throughout the continent, and extends its nesting area westwards to the Cape Verd Islands.

BREEDING HABITS : The Little Egret is a migrant to Europe, reaching its breeding grounds between the end of March and the beginning of May. It is gregarious and social, breeding not only in company with its own kind, but in that of various other species. Of its pairing habits I find nothing definite recorded, but it is possible that this and all other Herons pair for life, returning as they do to the old nesting places every year. The favourite breeding haunts of the Little Egret in Europe are the flooded willow and alder thickets in river valleys which are annually flooded, and in similar trees on the banks of lakes, or on bent down reeds, and in swamps. The nest, which is slight and shallow, is made of twigs, and these are arranged, as remarked by Mr. Seebohm, like spokes of a wheel from a common centre. He also observed that many of the twigs were green, and on some of them leaves still remained. The nests made in swamps are composed of dead reeds and other aquatic vegetation. This Egret becomes very noisy when disturbed at the nest, and rises wheeling into the air to join the crowds of other Herons that are flying about in alarm at the intrusion.

RANGE OF EGG COLOURATION AND MEASUREMENT: The eggs of the Little Egret are from three to six in number. They are chalky in texture, unpolished, and bluish-green in colour. Average measurement, 1·75 inch in length, by 1·3 inch in breadth. The duration of the period of incubation is unknown, as is also the sex that performs the duty.

DIAGNOSTIC CHARACTERS: There is no constant character by which the eggs of the Little Egret can be distinguished from those of the Night Heron; all that can be said is that on an average they are smaller.

Family ARDEIDÆ.　　　　　　　　　　Genus ARDEA.
GREAT WHITE EGRET.
ARDEA ALBA, *Linnæus*.
(British : Very rare abnormal spring and autumn migrant.)
Single Brooded. Laying season in Europe, May and June.

BREEDING AREA : Southern Palæarctic region. The typical form of the Great White Egret breeds in the valley of the Lower Danube, in Southern Russia, Asia Minor, Palestine, Persia, Turkestan, South-western Siberia (north to lat. 47), the Amoor valley, Manchooria, and Japan. Whether this Egret breeds anywhere in the Ethiopian region appears not to be known. Even its nesting in North Africa is very problematical.

BREEDING HABITS : The Great White Egret arrives at its European breeding grounds in April. It does not appear to be quite so gregarious as its allies, but probably this is due to the great decrease of this species owing to incessant persecution for the sake of its plumage. It seems, however, to be just as socially inclined, and often to breed in the company of other Herons, as well as Ibises and Pygmy Cormorants. Its favourite breeding haunts are inundated forests of low trees, and on the outskirts of dense thickets of reeds. Thirty years ago Von Homeyer found a solitary nest of this bird in Silesia, made in an old fir tree ; whilst in India the small race of this species (*Ardea alba*) habitually breeds on trees. The nest is flat, and platform-like, composed entirely of sticks, the finer twigs being used for the lining. Nests made in swamps are formed of dead reeds and other aquatic vegetation. It is probable that this Egret pairs for life, as it is said to return to its old breeding places, and to repair its old nests. The actions of this bird at the nest are not known to differ from those of allied species already described.

RANGE OF EGG COLOURATION AND MEASUREMENT: The eggs of the Great White Egret are from three to five in number, but the latter amount is exceptional. They are chalky in texture, with no polish, and greenish-blue in colour. Average measurement, 2·5 inches in length, by 1·5 inch in breadth. Incubation lasts about a month, but whether the male shares the duty with the female is unknown.

DIAGNOSTIC CHARACTERS: The eggs of the Great White Egret cannot be distinguished from those of the Common and Purple Herons, rendering their identification at the nest an imperative task, which is made easy by the fact of the present bird being very distinct in colouration.

Family ARDEIDÆ.   Genus ARDEA.

PURPLE HERON.

ARDEA PURPUREA, *Linnæus*.

(British: Rare abnormal spring and autumn migrant.)

Single Brooded. Laying season in Europe, April to June: varies greatly elsewhere.

BREEDING AREA: Southern Palæarctic region, Ethiopian and Oriental regions. The Purple Heron breeds in Central and Southern Germany, in Holland, France, the Spanish peninsula, Italy, Sicily, the valley of the Danube, and Southern Russia. It also breeds in Asia Minor and Palestine, Persia, Turkestan, India, Ceylon, and Burma. Whether it breeds in Borneo, Celebes, the Philippine Islands, Sumatra, Java, etc., appears not to be known, and the evidence for its nesting in China, Japan, and Southern Siberia is not conclusive. In Africa it appears to breed in suitable localities throughout the continent up to considerable elevations (Abyssinia, 9000 feet), and possibly also in Madagascar.

BREEDING HABITS: The Purple Heron reaches its European breeding haunts during March and April. It is not such a gregarious bird as the other species in this family previously described, and appears either to breed in very small colonies or in scattered pairs in the colonies of its relations. It is somewhat skulking in its habits. Of its pairing habits nothing appears to be known, but the bird probably mates for life like its allies. Its choice of a breeding ground is varied. In some localities the flooded forests of willow and alder and other trees are preferred ; in others dense reed beds are the attraction. When in a tree the nest is large and flat, and made of sticks, as it also is in some cases when built on crushed and broken reeds, as was remarked by Mr. Oates in Burma, who states that he found large colonies of this Heron. When amongst reeds it is generally supported on a bunch of these plants which have been crushed down into a kind of platform, and is composed of broken pieces of reed and other aquatic vegetation. When flushed from the nests the actions of this bird resemble those of allied species already described.

RANGE OF EGG COLOURATION AND MEASUREMENT: The eggs of the Purple Heron are from three to five in number, generally three. They are chalky in texture, unpolished, and greenish-blue. Average measurement, 2·2 inches in length, by 1·6 inch in breadth. The duration of the period of incubation is unknown. Whether the male takes any share in the duty appears not to have been determined.

DIAGNOSTIC CHARACTERS: The eggs of the Purple Heron cannot be distinguished from those of the Common Heron, and unless thoroughly well authenticated are of no scientific value. To a great extent the nesting habits of each species are different.

Family GRUIDÆ. Genus GRUS.

COMMON CRANE.

GRUS COMMUNIS, *Bechstein*.

(British : Formerly bred : Rare spring and autumn coasting migrant.)

Single Brooded. Laying season, from end of April to June, according to locality.

BREEDING AREA: Palæarctic region. The Common Crane breeds across Europe and Asia, from Scandinavia to Kamtschatka, in Europe reaching as far north as lat. 68°, but in West Siberia apparently not beyond the Arctic Circle, and in the far east no higher than Kamtschatka. It also breeds in Eastern Asia as far south as the valley of the Amoor, the Baikal area, and Russian Turkestan, whilst its southern limits in Europe include South Russia down to the Black Sea, Turkey, the Danube valley, Italy, and Andalucia in Spain, North Germany, Poland, Prussia, and Pomerania to the Elbe.

BREEDING HABITS: The migration of the Crane from its winter quarters to its breeding grounds commences as early as February and March, and continues through April into May. It is a gregarious bird in winter and on passage, but appears to disperse into scattered pairs for the nesting season. The Crane most probably pairs for life, and there is evidence to show that it returns year after year to one place to breed, and in many cases uses the same nest each season. The breeding haunts of this bird are swamps, near which open water occurs, either in the vicinity of forests, or on treeless steppes and tundras. The nest varies a good deal in size, the largest structures usually being in the most swampy situations, and the smaller nests on the comparatively dry hummocks or mounds in the swamps. If the nest is small it is little more than a trampled hollow, lined with dead leaves of sedges, or bits of withered broken rush. If the nest is large it is a heap of dead and half-rotten reeds, sedges, and other aquatic herbage, and branches of heath and twigs, as much as two feet across, the hollow holding the eggs being lined with the finest and driest material. The birds are wary enough at the nest, the male keeping watch near at hand, and the female slipping off the moment danger is detected.

RANGE OF EGG COLOURATION AND MEASUREMENT: The eggs of the Crane are almost invariably two in number, but very exceptionally three. They vary in ground colour from brownish-buff to greenish-buff, blotched and spotted with rich reddish-brown and pale brown, and with underlying markings of gray. The shell is rather rough in texture, pitted almost like tanned pigskin, and without polish. On some specimens the markings are mostly confluent and massed on the larger end; on others they form a zone round the larger end; others have the markings pretty evenly distributed over the entire surface, and pale and indistinct; whilst others have most of the surface colour suffused over the surface, here and there intermingled with very dark brown spots. Average measurement, 3·9 inches in length, by 2·5 inches in breadth. Incubation, apparently performed by the female, is said to last a month.

DIAGNOSTIC CHARACTERS: The large size is a reliable character, and serves to distinguish the eggs of the Common Crane from those of allied British species.

Family GRUID.E. Genus GRUS.

DEMOISELLE CRANE.

GRUS VIRGO (*Linnæus*).

(British: Very rare abnormal spring migrant.)

Single Brooded. Laying season from end of April to June and even July, according to locality and state of season.

BREEDING AREA: Southern Palæarctic region. In Europe the Demoiselle Crane breeds in Southern Spain, on the low-lying western shores of the Black Sea, and on the treeless steppes of Southern Russia, between the Caucasus and the fiftieth parallel of north latitude. In Asia it breeds in Turkestan, South-western Siberia (north to lat. 53°), Dauria, the Baikal area, Eastern Mongolia, and North-western China.

BREEDING HABITS: The Demoiselle Crane reaches its breeding grounds during March and April. During winter this Crane is gregarious, and even throughout the breeding season is social to a great extent, although the nests are scattered up and down the frequented district. There can be little doubt that this Crane also pairs for life, and returns regularly to certain places to breed. The favourite breeding haunts of this bird are in the vicinity of water on steppes, and vast sandy plains, little partiality being shown for swamps. The nest is always made upon the ground, either amongst grain or grass, or, according to

Dybowski, on the rocky banks of a river. It is a slight structure, a mere hollow trodden in the ground, and lined with a few bits of herbage. Nests described by Dybowski are said to have been made of small stones fitting close to each other, the surface being flat, and deepening towards the centre. I have found nests of the Fulmar very similar in construction. The parent birds are very wary at the nest, one being stationed near by to give the alarm of approaching danger, when the sitting bird slips quietly away, running or walking for a little distance, then taking flight. They are very pugnacious, and beat off any predaceous bird or animal with considerable courage.

RANGE OF EGG COLOURATION AND MEASUREMENT: The eggs of the Demoiselle Crane are two in number. They vary from pale buff to olive-brown in ground colour, spotted and blotched with umber-brown, and with underlying markings of lilac-gray. Two very distinct types occur, one in which the surface spots predominate, the other in which the underlying markings are most numerous and conspicuous. Average measurement, 3·5 inches in length, by 2·0 inches in breadth. Incubation, performed by both sexes, lasts about a month.

DIAGNOSTIC CHARACTERS: The eggs of the Demoiselle Crane cover much the same range of colour variation as those of the Common Crane, but may always be distinguished from those of that species by their much smaller size.

Family OTIDIDÆ. Genus OTIS.
GREAT BUSTARD.
OTIS TARDA, *Linnæus*.

(British : Formerly bred ; irregular nomadic spring, autumn, and winter migrant.)

Single Brooded. Laying season, May.

BREEDING AREA : Southern Palæarctic region. The Great Bustard breeds locally in Northern Prussia, Pomerania, Poland, Denmark, Spain, Italy, the steppes of the Danube and Turkey, and in Russia south of lat. 55 . Eastwards in Asia it breeds in Palestine, Turkestan, Siberia as far north as Omsk, the Baikal area, and the valley of the Amoor, southwards into Manchooria. It is not known with certainty whether this species still continues to breed in North-west Africa.

BREEDING HABITS: The migrations of the Great Bustard are limited to the northern portions of its breeding area. It is an early migrant, too, as is usual in this class of travellers, and appears in its summer haunts in the north as early as March or April. The favourite haunts of this Bustard are vast plains and steppes devoid for the most part of trees, and extensive grain lands. At all seasons this species is a social one, for during the breeding season several pairs of birds will meet and feed in company ; in winter the flocks become larger. Some observers have asserted that the Great Bustard is polygamous, but the evidence is by no means conclusive ; on the other hand, Naumann, who had abundant means of verifying his opinion, states that it pairs early in spring. In confirmation of this, it may be remarked that the sexes are said to keep in separate flocks during winter. The nest is always placed upon the ground, often in a bare situation on the open steppe, or in a field of young corn. It is merely a hollow trampled out by the female, about eighteen inches across, but not more than one or two inches deep, and lined with a few scraps of dry herbage. The bird is a close sitter, but a wary one, and runs from the nest as soon as danger threatens, taking wing a short distance away, and generally resorting to some remote spot, without demonstration of any kind.

RANGE OF EGG COLOURATION AND MEASUREMENT: The eggs of the Great Bustard are two or three in number, usually the former. They are coarse in texture, the surface full of minute pores, and vary from olive-green to olive-brown and pale buff in ground colour, spotted and blotched with reddish-brown, and with underlying markings of pale lilac-gray. The markings are seldom very bold or decided, and for the most part distributed here and there over the entire surface of the egg. On some specimens a few blackish-brown streaks occur. Average measurement, 3·0 inches in length, by 2·2 inches in breadth. Incubation, performed by the female, lasts from three weeks to a month.

DIAGNOSTIC CHARACTERS: The colour or size of these eggs readily distinguish them from those of allied species.

Family OTIDID.E. Genus OTIS.

LITTLE BUSTARD.

OTIS TETRAX, *Linnæus*.

(British : Rare nomadic spring, autumn, and winter migrant.)

Possibly Double Brooded. Laying season, May and July.

BREEDING AREA : South-western Palæarctic region. The Little Bustard breeds in suitable localities in the Spanish peninsula and in France, more commonly on the steppes of the Danube, Turkey, Southern Russia, and Western Siberia, as far north as lat. 55°, and as far east in the latter country as Lake Saisan. It also breeds in Russian Turkestan, Northern Persia, North-west Africa, and the islands of Sardinia and Sicily.

BREEDING HABITS : Although many Little Bustards remain on the northern shores of the Mediterranean during winter, the greater number cross that sea to winter in Africa, returning the following April. It is gregarious enough in winter and whilst on migration, but soon after the breeding grounds are reached the flocks disband, and not even a social tendency is observable until the young are abroad. The Little Bustard pairs annually. At this season numbers of birds congregate on certain spots, and the males appear to go through a sort of "lek," like many Game Birds, and fights take place for the possession of the females. As soon as pairing is over, each couple retire to a selected haunt until the young are grown. The favourite breeding grounds of this Bustard are wide treeless plains and steppes. The nest is invariably made upon the ground, and is a mere trampled hollow, seven or eight inches across, rather deep, and lined with a few bits of dry grass and weed. The birds are wary at the nest, the female slipping off at the approach of danger, usually being warned by the male, who keeps a close and constant watch in the neighbourhood. When flushed, the bird flies straight away without demonstration of anxiety or any alluring action.

RANGE OF EGG COLOURATION AND MEASUREMENT: The eggs of the Little Bustard are from three to five in number, four being the average clutch. They vary in ground colour from olive-brown to olive-green, and are indistinctly mottled with pale reddish-brown. The shell is polished and smooth, the pores being only slightly defined. As a rule, the larger end of the egg shows the most colouring, sometimes the smaller end. Average measurement, 2·0 inches in length, by 1·5 inch in breadth. Incubation is performed apparently entirely by the female, but the duration of the period is undetermined.

DIAGNOSTIC CHARACTERS: The eggs of the Little Bustard cannot readily be confused with those of any other western Palæarctic species, and are easily distinguished by their colour and size.

Family OTIDIDÆ. Genus OTIS.

MACQUEEN'S BUSTARD.

OTIS MACQUEENI, *Gray.*

(British: Very rare abnormal autumn migrant.)

Number of Broods unknown. Laying season, apparently unknown.

BREEDING AREA: South-central Palæarctic region. Macqueen's Bustard breeds in extreme south-western Siberia, as far east as Lake Saisan, in Turkestan and Persia, and southwards into Afghanistan, and possibly Arabia.

BREEDING HABITS : Of the habits of Macqueen's Bustard during the nesting season absolutely nothing appears to be known. It is a migratory bird, and arrives at its winter quarters in the Indian region in September, leaving them again for its northern breeding grounds in March or April. It is said to be monogamous, and pairs annually. Its breeding haunts are vast plains, apparently such as are dry, arid, or sandy being preferred. The nest of this bird has never been described.

RANGE OF EGG COLOURATION AND MEASUREMENT : The number of eggs laid by Macqueen's Bustard is unknown. Eggs of this bird obtained by Herr Tancré's collectors on the Altai Mountains are buffish or olive-brown in ground colour, blotched and spotted with rich dark brown and pale brown, and with underlying markings of dull gray. Average measurement, 2·55 inches in length, by 1·75 inch in breadth. It is impossible to say whether the two eggs purchased from an Arab at Aden in March by Lieut. Barnes belong to this species. These eggs were highly incubated at that date. The duration of the period of incubation is unknown, as is also which sex performs the duty.

DIAGNOSTIC CHARACTERS : The eggs of Macqueen's Bustard resemble very closely those of the Houbara Bustard, but are a trifle larger and darker in ground colour. Unless thoroughly well authenticated and with a reliable locality, the eggs of the present bird are of no scientific value, as the breeding areas of these two Bustards approach very near in Palestine. It will be remarked that the eggs of both these Bustards are more Plover-like in colour than those of the two preceding species, owing probably to the birds breeding on dry arid plains, instead of on grain or grass-covered steppes.

Family CHARADRIIDÆ.   Genus CURSORIUS.
Sub-family *CHARADRIINÆ*.

CREAM-COLOURED COURSER.

CURSORIUS GALLICUS (*Gmelin*).

(British : Rare abnormal spring and autumn migrant.)
Number of Broods unknown. Laying season, March to August, according to locality.

BREEDING AREA: South-western Palæarctic region, north-western Oriental region, and north-eastern Ethiopian region. The Cream-coloured Courser does not breed anywhere in Europe. It does so in the Canary Islands, and thence across Northern Africa from Morocco to Egypt southwards to Kordofan, and possibly Abyssinia. In Asia it breeds probably throughout Arabia and the trans-Caucasian steppes, in Persia, Afghanistan, Baluchistan, Rajputana, Scinde, and the Punjaub. It is possible that the race known as *C. gallicus bogolubovi* is the one predominating in some of the latter regions, and that *C. somalensis* may predominate in Arabia.

BREEDING HABITS: The Cream-coloured Courser can scarcely be called a migratory bird, except in the Caucasus. It is very bustard-like in its habits, frequenting vast arid plains and desert regions, where little or no cover occurs. It is one of the most characteristic birds of the Sahara, living amongst the sand-hills and dunes, either on the borders of the oases or far in the actual desert. The pairing habits of this bird almost exactly resemble those of the Great Bustard. It is monogamous, pairing every season, but as soon as the eggs are laid the males apparently flock by themselves until the young are hatched, when they rejoin their mates, and assist in rearing their offspring. In winter this Courser is even more gregarious, whilst in summer the flocks of immature non-breeding birds keep mostly together. The bird does not breed in colonies, but in scattered pairs. In the Canary Islands (where Mr. Meade Waldo [*Ibis*, 1893, p. 203] states that in 1891 about a thousand eggs of this bird were taken at Fuerteventura) and on the deserts of North Africa, the Cream-coloured Courser makes no nest, laying its eggs amongst small stones, or in a slight hollow in the sandy ground. In India, however, according to Hume, the nest is occasionally placed on stubble, or near a tuft of grass, under a bush, or amongst jungle, and is a hollow about five inches across and two inches deep, sometimes lined with a little dry grass. This different mode of nesting appears to have some effect on the colour of the eggs, those from India being much darker than those from Africa. The nests are extremely difficult to find, the bird slipping off at the first alarm, and going straight away, leaving the eggs to the safety which their protective colours ensure.

RANGE OF EGG COLOURATION AND MEASUREMENT: The eggs of the Cream-coloured Courser are usually two, and very rarely three in number. They are pale buff in ground colour, spotted, blotched, and freckled with buffish-brown, and marbled with underlying markings of gray. They are rotund in form, and smooth in texture. Average measurement, 1·2 inch in length, by 1·0 inch in breadth. Incubation is performed by the female, but the duration of the period is undetermined.

DIAGNOSTIC CHARACTERS: The size, form, and colour of the eggs of the Cream-coloured Courser prevent them being confused with those of any other European bird. We are not in possession of sufficient information and material to say whether the eggs of this Courser can be distinguished from those of the allied species, and if so, in what manner.

Family CHARADRIIDÆ. Genus GLAREOLA.
Sub-family CHARADRIINÆ.

COMMON PRATINCOLE.

GLAREOLA PRATINCOLA (*Linnæus*).

(British : Rare abnormal spring and autumn migrant.)

Single Brooded. Laying season, May.

BREEDING AREA : South-western Palæarctic region. The Common Pratincole breeds sparingly in Savoy, in France (principally in the south and central portions), Spain, the Balearic Islands, possibly the west coast of Italy, in Sicily, the lower valley of the Danube, Greece, Asia Minor, Palestine, Persia, and Russian Turkestan as far as Ala-Kul on the frontiers of Mongolia. South of the Mediterranean it breeds in North Africa, from Morocco eastwards to Tunis, and possibly further.

BREEDING HABITS : The Pratincole is a somewhat early migrant, reaching its breeding grounds in North Africa and South Europe in April. It is a gregarious bird during passage, but can scarcely be considered so at all of its breeding grounds, although many scattered nests may be found within a comparatively small area in some colonies, much closer together in others. Of the pairing habits of this bird nothing appears to be known. It probably mates for life, and returns season after season to certain places to breed. Its haunts are marshes, sandy plains, lagoons, and low flat islands, the latter being preferred wherever choice is possible. The Pratincole makes no nest, laying its eggs on the black, hard, sun-baked mud, without even a hollow to hold them. When their breeding grounds are invaded the Pratincoles are gregarious enough, and flocking together from all parts of the scattered colony become restless and noisy, and often indulge in various antics, even shamming death or broken limbs to lure intruders away. Before the eggs are actually laid the bird is very prone to these strange actions.

RANGE OF EGG COLOURATION AND MEASUREMENT: The eggs of the Pratincole are two or three in number, rarely four. They are very fragile, very oval, and without polish. They vary from buff or citron to pale gray in ground colour, spotted, blotched, and streaked with blackish-brown, and with underlying markings similar in character of grayish-brown. Usually the markings are generally distributed over the entire surface, but sometimes become most abundant at the larger end of the egg, and the gray underlying markings are numerous and clearly defined. Average measurement, 1·2 inch in length by ·9 inch in breadth. The duration of the period of incubation is unknown, as is also which sex performs the duty.

DIAGNOSTIC CHARACTERS: The eggs of the Pratincole cannot readily be confused with those of any other British species, but from those of allied species I am unable to give characters by which they may be separated. The locality is of some service in identifying them.

Family CHARADRIIDÆ.   Genus VANELLUS.
Sub-family CHARADRIINÆ.

SOCIABLE LAPWING.

VANELLUS GREGARIUS (*Pallas*).

(British: Very rare abnormal autumn migrant.)

Number of Broods unknown. Laying season unknown.

BREEDING AREA: South Central Palæarctic region. The Sociable Lapwing is believed to breed on the steppes of South-eastern Russia, from the Crimea as far north as Sarepta, and lat. 33°, and southwards to Astrakhan and the Caucasus. It is also presumed to breed in South-western Siberia and Turkestan as far east as Lake Saisan and Western Mongolia.

BREEDING HABITS: Of the habits of the Sociable Lapwing during the nesting season nothing whatever is known. It is said to frequent the steppes and plains for breeding purposes, but its nest has never been described.

RANGE OF EGG COLOURATION AND MEASUREMENT: All that is known respecting the eggs of the Sociable Lapwing is contained in Dresser's *Birds of Europe*. That naturalist writes—"A single egg sent to me by Mr. Möschler, who informs me that it was obtained by his Sarepta collector, with the birds, closely resembles eggs of the Common Lapwing (*Vanellus cristatus*), but is, if anything, rather paler in ground colour, and a trifle more sparingly marked with spots and blotches."

DIAGNOSTIC CHARACTERS: In the present state of our knowledge it is impossible to attempt to give any characters by which the eggs of the Sociable Lapwing may be distinguished from those of allied species.

Family CHARADRIIDÆ. Genus ÆGIALITIS.
Sub-family *CHARADRIINÆ*.

KILLDEER PLOVER.
ÆGIALITIS VOCIFERA (*Linnæus*).

(British : Very rare abnormal spring (?) and autumn migrant.)
Single Brooded. Laying season, April, May, and June, according to locality.

BREEDING AREA : Nearctic region, except the extreme north. The Killdeer Plover breeds sparingly in Mexico, and more generally throughout the United States northwards to Southern Canada, as far as about lat. 53°.

BREEDING HABITS : The Killdeer Plover is a migrant only in the colder portions of its range, but returns very early in spring as soon as its haunts are free from ice and snow. It is an inland bird, and its favourite breeding grounds are on the prairies on the banks of rivers and lakes, in more or less swampy districts. It is not exactly a gregarious bird during the breeding season, but numbers of pairs nest within a small area of suitable country, and are to a certain extent social in their habits. This Plover probably pairs annually. The nest is always made upon the ground, sometimes on bare sandy tracts, at others on ground studded with grass and rush tufts. It is merely a hollow in the ground scantily lined with a few bits of dry grass or other herbage, but in some cases even this slight provision is omitted. This Plover is wary at the nest, leaving its eggs at the first sign of danger, just like our own Lapwing, the male usually giving the alarm, flying towards the intruder and uttering its plaintive *too it*, and soon being joined by most of the other Plovers nesting in the vicinity. Sometimes alluring motions are employed.

RANGE OF EGG COLOURATION AND MEASUREMENT: The eggs of the Killdeer Plover are four in number. They are pear-shaped, smooth in texture, and pale buff of various shades in ground colour, blotched and spotted with blackish-brown and with underlying markings of brownish-gray. As a rule most of the blotches are on the larger end of the egg, forming a semi-confluent cap or an irregular zone. A not unfrequent type has the markings in the form of scratches and irregular spots. Average measurement, 1·6 inch in length, by 1·1 inch in breadth. Incubation is performed by both parents, but the duration of the period is unknown.

DIAGNOSTIC CHARACTERS: The eggs of the Killdeer cannot readily be confused with those of any other allied species breeding in the same area, their size and colouration being sufficient to distinguish them, with the sole exception of those of the Spotted Sandpiper, from which, however, their much larger size serves to separate them.

Family CHARADRIIDÆ. Genus ÆGIALITIS.
Sub-family CHARADRIINÆ.

RINGED PLOVER.

ÆGIALITIS HIATICULA (*Linnæus*).

(British: Possibly breeds; common spring and autumn coasting migrant.)

Single Brooded. Laying season, April to June, according to locality.

BREEDING AREA: Western Palæarctic region and north-eastern Neartic region. The typical form of the

Ringed Plover breeds in Cumberland Bay, on the American coast of Davis Strait, on the coasts of Greenland up to lat. 79°; in Iceland, Spitzbergen, Nova Zembla, and probably Franz-Josef Land. In Europe it breeds in suitable localities north of the Alps, and in Northern Africa, including the Canary Islands (although I notice Mr. Meade Waldo only records it as passing on migration) and Madeira. In Asia it is known to breed as far east as the Taimyr peninsula in the north, and the Baikal area in the south, in Turkestan, and Western Siberia. It is said also to breed on the coasts of the Red Sea, but confirmation is wanting.

BREEDING HABITS: This form of Ringed Plover is a migrant, and regularly passes the British coasts to those portions of its breeding area that lie to the north of our islands. A few pairs may possibly breed within our area, especially on the coasts of Kent and Essex. It arrives in the Arctic regions towards the end of May or early in June. Of its pairing habits nothing appears to be known, but the bird probably mates annually, the flocks disbanding at the breeding grounds. Although gregarious in winter it does not breed in colonies, but in more or less scattered pairs, which are certainly socially inclined. This Plover is not known to differ in its nesting habits from the larger race. It is rather remarkable, however, that Capt. Feilden found a nest of this form lined with the green fleshy leaves and stems of *Atriplex littoralis*, a fact which suggests, if it does not actually prove, some difference of habit. It may be that in the high north some sort of lining is added to the sandy nest for the purposes of warmth.

RANGE OF EGG COLOURATION AND MEASUREMENT: The eggs of the Ringed Plover are four in number, pyriform, and smooth in texture. They cover the same range of colour variation as those of the Greater Ringed

Plover. Average measurement, 1·3 inch in length, by ·9 inch in breadth. Incubation, performed by both sexes, lasts about three weeks.

DIAGNOSTIC CHARACTERS: The eggs of the Ringed Plover do not differ in colour from those of the Greater Ringed Plover, but are perceptibly smaller in size. The locality is of great importance in their identification.

Family CHARADRIIDÆ. Genus ÆGIALITIS.
Sub-family *CHARADRIINÆ*.

LITTLE RINGED PLOVER.

ÆGIALITIS MINOR (*Wolf and Meyer*).

(British : Rare abnormal spring and autumn migrant.)

Single Brooded. Laying season, May and first half of June.

BREEDING AREA: Palæarctic region. The Little Ringed Plover breeds throughout Europe and Palæarctic Asia south of lat. 60°, and south of the Mediterranean, in Africa north of the Great Desert, and possibly in the Canaries, although Mr. Meade Waldo does not record this species.

BREEDING HABITS: In the more northerly portions of its distribution the Little Ringed Plover is a migrant, retiring south to Africa in winter, and returning north again in April. This bird is not so gregarious as the Ringed Plovers, even in winter, but is to a certain extent social, numbers of pairs frequenting a comparatively small area. The favourite breeding haunts of this Plover are the sandy banks of rivers and inland lakes, low flat islands, and less frequently sand dunes and shingly beaches close to the sea. It pairs annually, and just about the period the male may be frequently seen soaring in the air uttering a by no means unmusical trill. I observed many pairs of this Plover in the fast drying-up bed of a river in the oasis of Biskra, and from time to time remarked the males careering about the air after they had reached the zenith of their flight. The Little Ringed Plover does not make any nest, the eggs being laid in a little hollow in the sand or shingle, no lining of any kind apparently ever being inserted. Mr. Abel Chapman states that he frequently found the eggs in a slight hollow scraped in dry cattle-droppings. The bird sits very lightly, and as it leaves the nest the moment danger threatens, the eggs are only found by a close search, as they resemble the ground around them. During hot sunny weather the eggs are often left uncovered for a considerable time, whilst the parent birds are feeding in the vicinity.

RANGE OF EGG COLOURATION AND MEASUREMENT: The eggs of the Little Ringed Plover are four in number. They are very pyriform in shape, buff in ground colour, speckled and streaked with various shades of brown, and with underlying markings of ink-gray. Most of the markings are on the larger end of the egg, but others are fairly well sprinkled over the surface. Average measurement, 1·15 inch in length, by ·85 inch in breadth. Incubation, performed by both sexes, lasts about three weeks.

DIAGNOSTIC CHARACTERS: The streaky character of the markings on the eggs of the Little Ringed Plover make them somewhat closely resemble those of the Kentish Plover, from which, however, they are readily distinguished by their smaller size.

Family CHARADRIIDÆ. Genus ÆGIALOPHILUS.
Sub-family CHARADRIINÆ.

CASPIAN SAND PLOVER.

ÆGIALOPHILUS ASIATICUS (*Pallas*).

(British: Very rare abnormal spring migrant.)

Number of Broods unknown. Laying season unknown.

BREEDING AREA: South central Palæarctic region. The Caspian Sand Plover is only presumed to breed in the basins of the Caspian and Aral seas. Its distribution during summer is most imperfectly known, and eggs, so far as I can learn, have been taken in two localities only.

BREEDING HABITS: The habits of the Caspian Sand Plover during the breeding season are practically unknown. It is a migratory bird, and in summer frequents sandy plains, and the banks and shores of rivers and seas. The nest appears never to have been described by any English naturalist.

RANGE OF EGG COLOURATION AND MEASUREMENT: The eggs of the Caspian Sand Plover are described by Mr. Dresser as oval and tapering in shape, warm buff with a faint tinge of green in ground colour, and the spots nearly black. This description appears to me a doubtful one. Middendorff figures an egg of this bird in his *Reise in Nord- und Ost-Sibirien*, vol. II. pl. xix. fig. 4.

DIAGNOSTIC CHARACTERS: In the present state of our knowledge it is impossible to say whether any character exists by which the eggs of this bird can be distinguished from those of allied species.

Family CHARADRIIDÆ. Genus CHARADRIUS.
Sub-family *CHARADRIINÆ.*

GRAY PLOVER.

CHARADRIUS HELVETICUS (*Brisson*).

(British: Common spring and autumn coasting migrant; few in winter.)

Single Brooded. Laying season, latter half of June, and early in July.

BREEDING AREA: Northern Nearctic and Palæarctic regions. The Gray Plover is only known to breed on the tundras above the limits of forest growth in the valley of the Petchora, on the Taimyr peninsula, and in the delta of the Lena; and in North America in Alaska, on the banks of the Anderson river, and on Melville peninsula.

BREEDING HABITS: The Gray Plover is a regular migrant to the Arctic regions, where it breeds, reaching them during the latter half of May or early in June. During winter and whilst on passage this Plover is gregarious, but at the breeding places the birds distribute themselves in scattered pairs, which nest in more or less close proximity, the males being social and often seen in small parties. The birds pair annually, and not apparently until they reach their destinations. The favourite breeding grounds of this Plover are the tundras and barren grounds between the limits of forest growth and the Arctic Ocean, especially such portions as are swampy and covered with ridges. Previous to 1875 the breeding habits of the Gray Plover were but little known, and eggs were very rare in collections. The first authentic eggs were those taken by Von Middendorff, on the Taimyr peninsula, in 1843. In 1864 Mac Farlane obtained eggs on the tundras near the Arctic

Ocean in North America. In 1875 Messrs Seebohm and Harvie-Brown discovered the first breeding grounds of this bird in Europe on the tundras of Northern Russia in the valley of the Petchora. Between June the 22nd and July the 12th, these two naturalists obtained ten nests, carefully identifying the parents at each. The nest is merely a slight hollow in the moss or lichen-covered ground, in which a few twigs, scraps of reindeer-moss, and other vegetable refuse are arranged. The birds were observed to indulge in rather curious flights as they rose from their nests. After being driven from the nest the female was usually the first to return, but she generally came less conspicuously than the male, making her appearance on a distant ridge of the tundra, then after looking round for a short time running quickly to the next ridge, and again looking round, calling at intervals to her mate with a single note. To this, however, the male seldom replied, but when he did so it was with a double note. After the female had run about thus for some time the male began to move, but he generally joined his mate by flying boldly up to her. On the other hand, the female rarely took to her wings. She was very cautious, and passed and repassed her nest several times until she finally settled upon it. All the time that the nest was being watched, the female was very restless, and ran about a good deal, but the male generally remained stationary on a hillock or ridge, apparently watching the movements of his mate.

RANGE OF EGG COLOURATION AND MEASUREMENT : The eggs of the Gray Plover are four in number. They are pyriform in shape, not quite so buff in ground colour as those of the Golden Plover, nor quite so olive as those of the Lapwing, spotted and blotched with blackish-brown and with underlying markings of gray. Several very distinct types occur. One is very sparingly marked, the spots being somewhat small, streaky, and irregular in shape, with one or two irregular lines; another has the blotches predominating, but small in size and mostly on the larger end of the egg; another is very heavily blotched, chiefly at the large end of the egg, most of them being confluent and joined by narrower makings; whilst another has large blotches, some of them confluent over most of the surface. Average measurement, 2·0 inches in length, by 1·4 inch in breadth. Incubation is apparently performed by both sexes, as MacFarlane records that a male was snared on one of the nests obtained by him. The duration of the period is unknown.

DIAGNOSTIC CHARACTERS: The eggs of the Gray Plover resemble those of the Lapwing in colour, and are about the same size, but the breeding grounds of the two species do not impinge, so that the locality is sufficient to identify them.

Family CHARADRIIDÆ.　　　　　Genus CHARADRIUS.
Sub-family CHARADRIINÆ.

ASIATIC GOLDEN PLOVER.

CHARADRIUS FULVUS, *Gmelin*.

(British : Very rare abnormal autumn migrant.)

Single Brooded. Laying season, end of June, or first half of July.

BREEDING AREA : North-eastern Palæarctic region. The Asiatic Golden Plover breeds on the tundras above the limits of forest growth of Eastern Siberia, from the valley of the Yenesay to the Pacific coast. It has been said to breed in New Caledonia.

BREEDING HABITS : The Asiatic Golden Plover is a migrant, and was observed to pass up the valley of the Yenesay to its breeding grounds in Siberia, during the first week in June, and doubtless reaches its summer haunts a week later. In winter and on migration it is gregarious, but our information of its habits during the breeding season is so meagre that we cannot say more than that it is to a certain extent social in summer, apparently nesting locally in scattered pairs over the ground. It probably pairs in spring, like its allies. The breeding haunts of this Plover are the tundras beyond the limits of the forests, where the ground is clothed more with moss and lichen than with grass, studded with patches of bare pebbly ground, and the dead flat relieved by hummocky plains. In such a locality the nest and eggs of the Asiatic Golden Plover were discovered by Mr. Seebohm during his visit to Siberia in the summer of 1877. On the 14th of July he observed a pair of these Plovers, and after much fruitless watching, one of them, the male, was shot. The nest was found shortly afterwards, amongst the moss and lichen, containing the full complement of eggs. It was merely a slight hollow lined with broken stalks of reindeer-moss. A week later the same naturalist found this Plover very common on the tundra at Golcheeka, and succeeded in obtaining the young in down. So far as I am aware these eggs are the only ones known to science. The behaviour of the old birds at the nest is very similar to that of allied species.

RANGE OF EGG COLOURATION AND MEASUREMENT : The eggs of the Asiatic Golden Plover are four in number. They are precisely similar to those of the Golden Plover, in colour and character of markings, varying in ground colour from light buff to very pale buff with a tinge of olive, blotched and spotted with rich dark brown, and with small underlying markings of gray. Average measurement, 1·9 inch in length, by 1·3 inch in breadth. The duration of the period of incubation is unknown, as is also whether the male takes any share in the task.

DIAGNOSTIC CHARACTERS: The eggs of the Asiatic Golden Plover very closely resemble those of the Golden Plover, but are smaller. The locality of the eggs also assists very considerably in their identification, as the European Golden Plover does not breed in the same area.

Family CHARADRIIDÆ. Genus CHARADRIUS.
Sub-family *CHARADRIINÆ*.

AMERICAN GOLDEN PLOVER.

CHARADRIUS FULVUS AMERICANUS (*Schlegel*).

(British. Very rare abnormal autumn migrant.)

Single Brooded. Laying season, end of June or first half of July.

BREEDING AREA: Northern Nearctic region. The American Golden Plover breeds on the barren grounds above the limits of forest growth in North America from Alaska to Greenland.

BREEDING HABITS: The American Golden Plover is a summer visitor to the barren grounds bordering the Arctic Ocean in the New World, reaching its nesting places at the end of May or in June. But little has been recorded of its habits during the breeding season, but as far as they are known they closely resemble those of the preceding race. MacFarlane met with the nest of this Plover. It was merely a hollow in the moss- or lichen-clothed ground, carelessly lined with a few scraps of herbage. The actions of the parents at the nest resemble those of allied species.

RANGE OF EGG COLOURATION AND MEASUREMENT: The eggs of the American Golden Plover are four in number. MacFarlane, who obtained 170 nests, records a clutch of *five*. They so closely resemble those of the Asiatic form that a description is unnecessary. MacFarlane gives the average measurements as precisely the same as those of the Old World bird.

DIAGNOSTIC CHARACTERS: The eggs of this race of Golden Plover can be safely identified by the locality from which they come.

Family CHARADRIIDÆ.   Genus HIMANTOPUS.
Sub-family TOTANINÆ.

COMMON STILT.

HIMANTOPUS MELANOPTERUS, *Meyer*.

(British. Rare abnormal spring and autumn migrant, chiefly the former.)

Single Brooded. Laying season, from end of April to June, according to locality.

BREEDING AREA: Southern and Western Palæarctic region, Ethiopian and Oriental regions. The Common Stilt breeds in the marshes of the Spanish peninsula, in the Delta of the Rhone, in Sicily, the valley of the Danube, and on the Lagoons of the Black Sea. Eastwards it breeds on the Kirghiz and Kalmuk Steppes, in Asia Minor, Palestine, Northern Persia, Turkestan, Afghanistan, India, and Ceylon. South of the Mediterranean it breeds locally throughout the African continent as far south as the Berg river, and probably in Madagascar.

BREEDING HABITS: The Common Stilt is with a few exceptions a summer migrant to Europe, and to such of its breeding grounds in Asia as are situated in northern districts. The bird is gregarious, and breeds in colonies of varying size, some consisting only of a few pairs, others of hundreds of pairs. The favourite haunts of this Stilt during the nesting season are salt marshes, the shores of lagoons, and low mud islands and banks. The nests are made in a great variety of situations, and vary considerably both in size and material. If the ground is wet the nest is bulkier than when on a dry site. Some nests are made absolutely in the water, heaps of dead reeds and other aquatic vegetation, rising several inches above the level of the water; others are made on the mud, and are mere hollows lined with dry grass, broken reeds, and bits of rush-leaf. A most interesting breeding place of this Stilt is situated at some salt works near Delhi, in Upper India. These works consist of many acres of shallow lime-lined pools, divided from each other by strips of ground, from one to six feet in width. On these narrow strips and in the shallow pools the nests are made. They are very curious structures, little platforms made of bits of lime, raised about three inches high, and from seven to twelve inches across, on which a slight bed of dry grass is placed. Many nests are made close together, and the birds, through being left unmolested, are remarkably tame, allowing the workmen to pass them closely as they sit upon their eggs. When disturbed at the colony, the birds rise from their eggs, or run from them with elevated wings before taking flight, becoming very noisy and anxious for the safety of their nests.

RANGE OF EGG COLOURATION AND MEASUREMENT: The eggs of the Common Stilt are four in number. They are pyriform in shape and pale or dark buffish-brown in ground colour, streaked, spotted, and blotched with blackish-brown of various shades, and with underlying markings of gray. Much variation occurs in the character of the marking. Some eggs are covered with moderately large and irregular blotches, some of the latter joined together with irregular streaky marks; others have the markings small and streaky; others have them large and pale. As a rule most of the spotting is on the major half of the egg, and the gray underlying markings are seldom conspicuous or numerous. Average measurement, 1·7 inch in length, by 1·2 inch in breadth. The duration of the period of incubation is unknown, neither has the sex which performs the task been determined.

DIAGNOSTIC CHARACTERS: Many of the eggs of the Common Stilt closely resemble those of the Avocet in colour, but are much smaller in size. I cannot find any reliable character by which the eggs of this Stilt can be distinguished from those of certain allied species, but the locality is sufficient to identify them.

Family CHARADRIIDÆ.   Genus RECURVIROSTRA.
Sub-family *TOTANINÆ*.

COMMON AVOCET.

RECURVIROSTRA AVOCETTA, *Linnæus*.

(British : Formerly bred ; rare spring and autumn coasting migrant.)

Single Brooded. Laying season, May, and early June.

BREEDING AREA : Southern Palæarctic and Ethiopian regions. The Common Avocet breeds on some of the islands off the Dutch and Danish coasts, on the marshes in the delta of the Rhone, in the marismas of Southern Spain, in the valley of the Danube, and on the lagoons of the Black Sea. Eastwards it breeds in Palestine, Persia, Northern Turkestan, South-western Siberia, South-eastern Mongolia, and Southern Dauria. South of the Mediterranean it is said to breed in all suitable localities throughout Continental Africa, and probably in Madagascar.

BREEDING HABITS: To Europe and to the more northern breeding grounds in Asia the Avocet is only a summer visitor, arriving from its winter quarters in April or May. At all seasons it is gregarious, and breeds in colonies of varying size. It probably pairs for life, as every season the same nesting places appear to be visited. The favourite haunts of the Avocet in summer are low sandy coasts, salt marshes, lagoons, and mud islands. The nests are either placed on the bare sand or mud, or amongst the short herbage of marshes, and are mere hollows lined with a few bits of dry grass or dead leaves. The Avocet is not a close sitter, leaving its eggs at the first sign of danger, and flying to and fro above the colony; it is ever ready, however, to try and drive off any intruding bird which might possibly have evil designs upon its eggs.

RANGE OF EGG COLOURATION AND MEASUREMENT: The eggs of the Avocet are generally three or four in number, but five or even six have been found in very exceptional cases. They are pyriform, smooth in texture, and pale buff in ground colour, spotted and blotched with blackish-brown, and with underlying markings of gray. Generally the eggs of the Avocet are covered on the large end with irregular and often confluent blotches, but a type occurs in which the spots are smaller and distributed over most of the surface. Average measurement, 1·95 inch in length, by 1·4 inch in breadth. Incubation, performed by both sexes, lasts, according to Naumann, seventeen to eighteen days.

DIAGNOSTIC CHARACTERS: Due allowance being made for *locality*, the only eggs likely to be confused with those of the Avocet are those of the Lapwing, but from these their lighter appearance and less bold markings are sufficient to distinguish them: they never also show any olive tint in the ground colour. Eggs of the Gray Plover also approach them in general appearance, but the breeding areas of the two species are quite distinct.

Family CHARADRIIDÆ. Genus NUMENIUS.
Sub-family *TOTANINÆ*.

ESKIMO WHIMBREL.

NUMENIUS BOREALIS (*J. R. Forster*).

(British: Very rare abnormal autumn migrant.)

Number of broods unknown. Laying season, June and early July.

BREEDING AREA: Northern Nearctic region. The Eskimo Whimbrel breeds on the barren grounds above the limits of forest growth of Arctic America from Alaska to the shores of Hudson Bay and Davis Strait.

BREEDING HABITS: The Eskimo Whimbrel is a regular migrant, passing the United States in May, but not arriving at its summer quarters until the ground is free from snow in June. On migration it is gregarious, but the flocks disperse into scattered pairs at the breeding grounds. This Whimbrel pairs annually, but of its habits at that period I find nothing recorded. The favourite nesting haunts of the Eskimo Whimbrel are in the marshy portions of the barren grounds apparently at no great distance from the shores of the Arctic Ocean.

The nest is always made upon the ground, and is a hollow scantily lined with a few scraps of dry herbage and withered leaves. Although MacFarlane, Richardson, and other naturalists have had ample opportunities of studying the habits of this bird during the breeding season, but little has been recorded, and many interesting points remain still undetermined. MacFarlane states that great difficulty was frequently experienced in finding the nest of this species, the eggs closely resembling surrounding objects in colour, and the hen gliding off her charge before being closely approached. He describes the note at the breeding grounds as a "prolonged mellow whistle."

RANGE OF EGG COLOURATION AND MEASUREMENT: The eggs of the Eskimo Whimbrel are four in number, although Richardson states that he has seen a female sitting on *three*—probably the fourth was about to be laid. They are pyriform, and range from brownish- or grayish-buff to greenish-olive in ground colour, blotched and spotted with brown of various shades, and with underlying markings of pale grayish-brown. The surface markings are generally large and boldly defined, and are chocolate or sepia-brown in colour, most numerous at the larger end of the egg, often confluent and sometimes forming an irregular cap. The gray underlying markings are similar in character and fairly numerous. Average measurement, 2·0 inches in length, by 1·4 inch in breadth. Incubation is performed chiefly by the female, but the duration of the period is unknown.

DIAGNOSTIC CHARACTERS: The eggs of this Whimbrel cannot readily be confused with those of any other species, the size, colour, and locality being sufficient to identify them.

Family CHARADRIIDÆ. Genus PHALAROPUS.
Sub-family TOTANINÆ.

GRAY PHALAROPE.

PHALAROPUS FULICARIUS (*Linnæus*).

(British : Rare nomadic autumn and winter migrant ; few in spring.)

Single Brooded. Laying season, first half of June.

BREEDING AREA : Northern Nearctic and Palæarctic regions. The Gray Phalarope is not known to breed anywhere in continental Europe. It breeds locally on the islands and coasts of Arctic Asia and America, reaching at least as far north as lat. $82\frac{1}{2}°$, and probably extending as high as land exists. It also nests in Iceland and Spitzbergen, on the Taimyr peninsula, in the delta of the Lena, the Tchuski Land, Alaska, the Parry Islands, Grinnell Land, and Greenland.

BREEDING HABITS : The Gray Phalarope is a nomadic migrant, wintering as far north as it can with safety, and returning to its breeding grounds late in May and early in June. The pairing habits of the Phalaropes are very abnormal, the females conducting the courtship, and leaving the males to incubate the eggs! The birds pair annually shortly after arriving in their summer quarters. This Phalarope is gregarious, winter and summer alike, and breeds in colonies of varying size dependent upon local conditions. The favourite breeding haunts of the Gray Phalarope are the marshy pools and lakes on the tundras at no great distance from the Arctic Ocean. The nest is made upon the ground close to the pools, and is merely a hollow in the moss or lichen, sometimes lined with a few dry leaves, or bits of grass. In their actions at the nests, when disturbed by man, these birds are not known to differ from allied species, being very tame and confiding, and often displaying alluring antics.

RANGE OF EGG COLOURATION AND MEASUREMENT: The eggs of the Gray Phalarope are four in number. They are pyriform in shape, smooth in texture, and pale buff with an olive tinge in ground colour, blotched and spotted with rich dark brown, and with underlying markings of pale brown. The surface markings are generally large and boldly defined, the underlying ones few and indistinct. As usual most of the blotches are on the larger end of the egg, and for the most part confluent. Average measurement, 1·25 inch in length, by ·87 inch in breadth. Incubation is performed by the male, but the duration of the period is unknown: it is probably about three weeks.

DIAGNOSTIC CHARACTERS: The eggs of the Gray Phalarope may be distinguished from those of the Red-necked Phalarope by their much larger size. They closely resemble those of Wilson's Phalarope; but the breeding areas of the two species are distinct, the *locality* of the eggs being quite sufficient to identify them.

Family CHARADRIIDÆ. Genus TOTANUS.
Sub-family TOTANINÆ.

BARTRAM'S SANDPIPER.

TOTANUS BARTRAMI (*Wilson*).

(British : Very rare abnormal autumn migrant.)

Single Brooded. Laying season, latter half of May and early June.

BREEDING AREA: Central Nearctic region. Bartram's Sandpiper breeds in Alaska, Rupert's Land, Southern Canada, and the more northerly United States, from Pennsylvania westwards to the Rocky Mountains.

BREEDING HABITS: Bartram's Sandpiper migrates north in April and May, crossing the prairies in vast numbers as well as following the coasts. For the greater part of the year this Sandpiper is very gregarious, and even in the breeding season numbers of pairs frequently nest close together, resembling the Lapwing in this respect. The favourite haunts of Bartram's Sandpiper are the open prairies and uplands, especially such as are studded with swamps and open pools: marshy places in wooded districts, where the timber is more or less scattered, are also frequent resorts. The pairing habits of this bird closely resemble those of other Sandpipers. The bird, at that season, may be seen perching on trees and fences, running along the branches or rails with uplifted wings, and uttering a long tremulous note. The nest is always made upon the ground amongst the herbage, and is merely a hollow lined with a little dry grass or a few dead leaves. The bird is a somewhat close sitter, and when alarmed usually runs from the nest, being apparently averse to flying until actually compelled. Very often various alluring antics are performed to endeavour to decoy an intruder from the nest. Many eggs of this species are collected for food.

RANGE OF EGG COLOURATION AND MEASUREMENT: The eggs of Bartram's Sandpiper are four in number. They are not quite so pear-shaped as usual, smooth in texture, and vary from pale grayish-buff to pale buffish-brown in ground colour, spotted and blotched with reddish-brown, and with underlying markings of gray. The markings are not large, and the surface spots are more or less circular, becoming most numerous on the larger end of the egg, where occasionally a few fine streaks of dark brown occur. Average measurement, 1·8 inch in length, by 1·3 inch in breadth. Incubation is performed by the female, but the duration of the period is not determined.

DIAGNOSTIC CHARACTERS: The eggs of Bartram's Sandpiper are readily distinguished by their size and colouration from those of all other allied species in the Nearctic region.

Family CHARADRIIDÆ.   Genus TOTANUS.
Sub-family *TOTANINÆ*.

SPOTTED SANDPIPER.

TOTANUS MACULARIUS (*Linnæus*).

(British: Rare abnormal spring and autumn migrant.)

Single Brooded.   Laying season, June.

BREEDING AREA: Nearctic region. The Spotted Sandpiper breeds in all suitable districts throughout the United States and British North America, as far north as about lat. 60°.

BREEDING AREA : The Spotted Sandpiper is a regular bird of passage, arriving in its more southerly breeding quarters in April, but a little later in the extreme northern limits of its distribution. As might naturally be expected, the habits of this species very closely resemble those of the Common Sandpiper, but American ornithologists do not appear to know whether the bird pairs for life, as its Palæarctic representative most certainly does. The favourite breeding places of the Spotted Sandpiper are the banks of rivers and the shores of lakes, either in wooded districts or on the open prairies, where only the scattered "bluffs" relieve the monotony of the otherwise treeless expanse. The nest is made upon the ground, and is merely a hollow lined with a few bits of dry grass, pine needles, or small dead leaves. Audubon states that in Labrador a much more elaborate structure is made, consisting of moss, grass, and feathers! but with all my respect for the great naturalist's observations, I cannot accept them unsubstantiated by the experience of other observers. The actions of this bird at the nest are similar to those of the Old World species.

RANGE OF EGG COLOURATION AND MEASUREMENT : The eggs of the Spotted Sandpiper are four in number. They vary in ground colour from creamy-white to rich buff, the paler eggs, according to Mr. Raine, being from the shores of alkaline lakes, and in a series very apparent. They are spotted and more rarely blotched with very dark reddish-brown, and with underlying markings of pale gray. As a rule the spots are small and circular, and the blotches are never very large. Most of the markings are generally on the larger end of the egg, but varieties are common in which they are pretty equally distributed over the whole surface. Average measurement, 1·3 inch in length, by 1·0 inch in breadth. Incubation is probably performed by both sexes, and lasts about three weeks.

DIAGNOSTIC CHARACTERS: The eggs of the Spotted Sandpiper are distinguished from those of allied species by their colour and size. They cannot be easily confused with those of the Common Sandpiper, but somewhat resemble those of the Killdeer Plover, although they are perceptibly smaller and blunter.

Family CHARADRIIDÆ.　　　　Genus TOTANUS.
Sub-family TOTANINÆ.

GREEN SANDPIPER.

TOTANUS OCHROPUS (*Linnæus*).

(British : Spring and autumn coasting migrant ; some, winter.)

Single Brooded.　Laying season, middle of April to end of June, according to locality.

BREEDING AREA: Northern and Central Palæarctic region. The Green Sandpiper breeds in the swampy forests about the vicinity of the Arctic Circle across Europe and Asia, and in a similar climate at high elevations in more southerly latitudes, in the Pyrenees, the Alps, the Carpathians, and the Caucasus, and eastwards on the mountains of Turkestan and Southern Siberia.

BREEDING HABITS: The Green Sandpiper reaches its southern breeding grounds in April, but further north it is from one to two months later. It can scarcely be classed as gregarious, nor is it at all social, even during winter. At its breeding grounds it lives in scattered pairs, which appear to keep exclusively to themselves.

Of the pairing habits of this Sandpiper nothing appears to have been recorded. With the possible exception of the Solitary Sandpiper, the breeding habits of the present species are, so far as is known, unique. Instead of making its nest upon the ground, it lays its eggs in trees, usually in the deserted home of some other bird. This extraordinary fact had long been known to some continental naturalists (although apparently the great German bird man, Naumann, was unaware of it), but was not generally known to British ornithologists until Professor Newton brought the circumstances before the Zoological Society of London, his communication being published in the *Proceedings* of that body in 1863. During the breeding season the Green Sandpiper is seen as often in trees and bushes as upon the ground. The favourite breeding haunts of this bird are open, marshy forests, the banks of wooded streams, and swampy thickets. A deserted nest of a Blackbird or a Thrush, a Jay or a Ring Dove, or even a Crow is often selected. As a rule old nests are preferred from three to twelve feet from the ground, but the eggs have been taken from an old drey of a squirrel as much as thirty feet above it, whilst others have been found in a hole in a fallen tree, and on the stump of a tree which had either been felled or blown down. Less frequently the eggs are laid in the hollow of a forking branch, on a heap of drifted leaves, or on lichen. Almost invariably the selected spot is close to water, and often in marshes.

RANGE OF EGG COLOURATION AND MEASUREMENT : The eggs of the Green Sandpiper are four in number. As *seven* have been found together, it would appear that two females sometimes agree to share the same spot. They vary from creamy-white, sometimes tinged with pale olive, to pale buff in ground colour, spotted with dark reddish-brown, and with underlying markings of pale grayish-brown. The markings are seldom large and blotchy, as is so characteristic in this group of birds, and are, of course, most numerous on the larger end of the egg. Average measurement, 1·55 inch in length, by 1·1 inch in breadth. The duration of the period of incubation is unknown. Whether the male assists in the duty is also undetermined.

DIAGNOSTIC CHARACTERS: The eggs of this Sandpiper perhaps approach most closely to those of the Common Sandpiper, but the markings are generally smaller. The situation of the "nest," however, is quite sufficient to identify them.

Family CHARADRIIDÆ. Genus TOTANUS.
Sub-family *TOTANINÆ*.

YELLOW-LEGGED SANDPIPER.

TOTANUS FLAVIPES (*Gmelin*).

(British: Very rare abnormal autumn migrant.)

Single Brooded. Laying season, end of May and early June.

BREEDING AREA: Northern Nearctic region. The Yellow-legged Sandpiper breeds across the North American continent from the Yukon valley in Alaska, in the west, to the Hudson Bay Territory and Greenland in the east, extending as far south as about lat. 44°.

BREEDING HABITS: The Yellow-legged Sandpiper is a migrant, arriving at its more southerly breeding grounds in May, but a month later in the extreme northern portions of its distribution. This species is not gregarious during the nesting season, but lives in scattered pairs. Of its pairing habits nothing appears to have been observed. The bird probably mates in spring, and carries on part of its courtship in the air accompanied by trilling notes, like so many other allied species are known to do. The favourite breeding grounds of this Sandpiper are open marshy districts, especially such as are studded with lakes, and the banks of rivers. The nest, which is made upon the ground, is often under the shelter of a bush or tussock of sedge. It is merely a small hollow, sometimes, but not always, lined with a few dead leaves and twigs. MacFarlane noticed on several occasions the male bird perch on a tree near the nest, and remarks its noisy habits.

RANGE OF EGG COLOURATION AND MEASUREMENT: The eggs of the Yellow-legged Sandpiper are four in number, pear-shaped, and smooth in texture. I have examined but few specimens of the eggs of this bird, but Mr. Seebohm, who has seen the remarkably fine series in the Smithsonian Institution at Washington, reports that the ground colour varies "from creamy-white to grayish-brown. The surface spots are dark rich reddish-brown, and vary in size from a large pea downwards, many of them becoming confluent and forming large irregular blotches, or occasionally taking the form of streaks. Most of the markings are generally on the larger end of the egg, but on some specimens they are more evenly distributed over the entire surface. The underlying markings are pale gray, or grayish-brown, and are large and conspicuous." Average measurement, 1·65 inch in length, by 1·1 inch in breadth. The duration of the period of incubation is unknown, as is also which parent performs the task: doubtless it is the female.

DIAGNOSTIC CHARACTERS: The eggs of this Sandpiper cover much the same range of colour variation as those of the Greenshank, but are distinguished from those of allied species by their size and somewhat narrow shape.

Family CHARADRIIDÆ. Genus TOTANUS.
Sub-family TOTANINÆ.

DUSKY REDSHANK.

TOTANUS FUSCUS (*Linnæus*).

(British : Rare spring and autumn coasting migrant ; some, winter.)

Single Brooded. Laying season, latter end of May, or early in June.

BREEDING AREA : Northern Palæarctic region. The Dusky Redshank breeds on the tundras above the limits of forest growth (but nowhere apparently south of the Arctic Circle unless on the highlands of Turkestan) in Arctic Europe and Asia, from Lapland in the west to the Tchuski Land in the east.

BREEDING HABITS : The Dusky Redshank begins to cross the Mediterranean in March, and continues to do so until the middle of May, reaching its summer quarters as soon as the ground is clear of snow. Except on migration this bird is not very gregarious, and when nesting is only found in widely scattered pairs. Of the pairing habits of this bird nothing has been recorded ; whilst of its habits during the nesting season generally we possess only the interesting observations made by Wolley in Lapland. He states that the favourite nesting grounds of this species were in the open parts of the forests, not necessarily near water, and especially in places where the trees had been burnt and the vegetation was scanty. He found the nests generally on rising ground, near the tops of hills, in open clearings amongst the pines where the earth was clothed with such plants as heath and reindeer-moss. They were mere hollows in the ground, lined with a few dead "needles" of the

Scotch fir. Wolley remarked that the bird sat closely, although its white rump was very conspicuous as it brooded over the eggs with its long neck drawn in. When flushed it either ran for a little way before taking wing, or flew into the air at once, and wheeled round and round uttering its note at intervals ; but sometimes it perched on the top of a tree near by. Although it sits so closely, it is described as being very wary in returning to the nest.

RANGE OF EGG COLOURATION AND MEASUREMENT: The eggs of the Dusky Redshank are four in number. They are pyriform in shape, smooth in texture, and vary from pale brown to pale green in ground colour, handsomely blotched and spotted with rich dark brown, and with underlying markings of pale brown and ink-gray. On many eggs a few bold hair-like lines occur. The eggs of this species are very boldly marked, often so much so as to cover the greater portion of the large end with a nearly confluent cap. Average measurement, 1·85 inch in length, by 1·3 inch in breadth. The duration of the period of incubation is unknown, as is also which sex performs the task.

DIAGNOSTIC CHARACTERS: The eggs of the Dusky Redshank so closely resemble those of the Great Snipe, that I can give no reliable character by which they may be distinguished. The markings are not so obliquely distributed as on those of the Snipe, and the ground colour of those of the latter are rarely green.

Family CHARADRIIDÆ. Genus LIMOSA.
Sub-family TOTANINÆ.

BLACK-TAILED GODWIT.

LIMOSA MELANURA, *Leisler.*

(British : Formerly bred ; regular spring and autumn coasting migrant.)

Single Brooded. Laying season, May and June, according to locality.

BREEDING AREA : Western Palæarctic region. The Black-tailed Godwit breeds in Iceland and the Faroes. On Continental Europe it breeds in Belgium, Holland, Denmark, Scandinavia (occasionally as far north even as the Arctic Circle), Central and Southern Russia, Poland, and Northern Germany. In Asia it breeds in South-west Siberia as far north as lat. 60°, and as far east as the western tributaries of the Obb, and in Western Turkestan.

BREEDING HABITS : The Black-tailed Godwit begins to leave its winter quarters in Africa in February, and continues to move north until the middle of March ; it travels somewhat slowly, and does not reach its breeding grounds until April or May. It is gregarious during winter and whilst on passage, but the flocks disband at the nesting places and scatter themselves in pairs over their haunts, many nests often being made, however, within a small area of suitable ground. The favourite breeding places of this Godwit are marshes and swampy meadows. The nest, invariably made upon the ground, is usually well concealed amongst the herbage, and is often placed in a tussock of sedgy grass. A dry spot in the marsh is generally selected. The nest is merely a hollow about three inches deep, but rather neatly lined with dry grass and other vegetable refuse. As soon as their breeding haunts are invaded, the old birds rise and fly to and fro with noisy clamour, rarely, if ever, remaining on their nests until closely approached.

RANGE OF EGG COLOURATION AND MEASUREMENT: The eggs of the Black-tailed Godwit are four in number. They are pyriform, smooth in texture, and vary from olive-brown to olive-green in ground colour, spotted and blotched with darker olive-brown, and with underlying markings of pale brown and gray. The markings as a rule are somewhat pale and ill-defined, but types occur in which they are bolder and clearer. Average measurement, 2·15 inches in length, by 1·5 inch in breadth. Incubation is performed chiefly by the female, but the duration of the period is unknown.

DIAGNOSTIC CHARACTERS: There is no constant character by which the eggs of this Godwit can be distinguished from those of the Bar-tailed Godwit. As the latter are far and away the rarest, great care should be exercised in their identification. The breeding range of the two species does not impinge, that of the Bar-tailed Godwit being above the limits of forest growth in the Arctic regions.

Family CHARADRIIDÆ.   Genus LIMOSA.
Sub-family TOTANINÆ.

BAR-TAILED GODWIT.

LIMOSA RUFA, *Brisson*.

(British : Common spring and autumn migrant ; some, late in winter.)

Double Brooded. Laying season, late in May, and early June.

BREEDING AREA : Northern and Western Palæarctic region. The Bar-tailed Godwit apparently breeds on the tundras above the limits of forest growth in Arctic Europe and Asia, from Lapland in the west to the valley of the Yenesay in the east. As yet it has only been observed in Lapland actually breeding.

BREEDING HABITS : The Bar-tailed Godwit begins to leave its winter quarters in Northern Africa in February, and the stream of migrants is slowly percolating northwards into Europe from that date until the end of April, but the breeding grounds on the Arctic tundras are not reached until the end of May or early in June. During winter this bird is gregarious, as it also is whilst on passage. During the pairing season in spring the male utters a trill whilst in the air. Of the actual breeding habits of this Godwit we possess but little information. Its breeding grounds are known to be the swampy moors of the Arctic regions, but the bird seems very locally distributed over them. Wolley obtained eggs in Finland, and he states that it breeds in marshes, the nest being very hard to find. The nest is described as merely a hollow, in some dry spot, lined with a little dead vegetable refuse. The bird's actions at the nest are not known to differ from those of allied species.

RANGE OF EGG COLOURATION AND MEASUREMENT : The eggs of the Bar-tailed Godwit are four in number.

They are pyriform in shape, olive-green of various shades in ground colour, spotted and blotched with darker brown, and with underlying markings of gray. In Mr. Seebohm's collection there are two eggs presumed to be of this species, which are very boldly marked, one specimen being also streaked with very dark brown on the larger end. Average measurement, 2·2 inches in length, by 1·45 inch in breadth. The duration of the period of incubation is unknown, as is also which parent performs the task.

DIAGNOSTIC CHARACTERS: The eggs of this Godwit cannot be distinguished from those of the Black-tailed Godwit. But very few genuine eggs of this bird are known in collections, and the student is warned against possible and easy fraud. The breeding area of the two species, however, is distinct. It may also be remarked that the eggs of this and the preceding species somewhat closely resemble those of Buffon's Skua—an additional reason for their careful identification.

Family CHARADRIIDÆ. Genus EREUNETES.
Sub-family TOTANINÆ.

RED-BREASTED SNIPE.

EREUNETES GRISEUS (*Gmelin*).

(British : Rare abnormal autumn migrant.)

Single Brooded. Laying season, June.

BREEDING AREA: North central Nearctic region. The Red-breasted Snipe breeds throughout the Arctic regions of North America from the Rocky Mountains in the west, to Baffin Bay in the east, south to Hudson Bay, and probably the Great Lakes, in about lat. 44°.

S

BREEDING HABITS: The Red-breasted Snipe is a somewhat late migrant, as is usual with strictly boreal species, not reaching its breeding grounds until the end of May or early in June, even later in the extreme north. During winter and whilst on migration this bird is gregarious, but in summer it lives in more or less scattered pairs, the immature non-breeding birds however remaining in flocks, some distance south of the nesting grounds. Of the pairing habits of this bird nothing appears to be known. The nesting grounds of the Red-breasted Snipe are situated on the tundras, or as the Americans call them "barren grounds," in swamps and near to lakes. Sometimes it breeds close to the sea, at others considerable distances inland. The nest is always made upon the ground, often in a tuft of marsh grass, or amongst the short vegetation on the shores of the moorland lakes. It is a mere hollow, scantily lined with a few dead leaves or bits of withered herbage. One of the birds which MacFarlane flushed from the nest rose for some height into the air, uttering a long shrill note of alarm. The bird is a somewhat close sitter, resembling the Common Snipe in this respect.

RANGE OF EGG COLOURATION AND MEASUREMENT: The eggs of the Red-breasted Snipe are four in number. They are pyriform, and smooth in texture, and vary in ground colour from pale greenish-brown to pale buffish-brown, blotched and spotted with dark reddish-brown and with underlying markings of pale grayish-brown. Sometimes a few very dark streaks occur. These eggs are very handsome ones, most of the big blotches being confluent and congregated on or near the large end. As in eggs of the Common Snipe, many of the blotches are obliquely distributed. Average measurement, 1·7 inch in length, by 1·15 inch in breadth. Incubation is performed apparently by the female, but the duration of the period is unknown.

DIAGNOSTIC CHARACTERS: There is no character by which the eggs of this bird can be distinguished from those of the Common Snipe and its Nearctic representative, Wilson's Snipe. To a certain extent the latter bird does not breed within the same area, not north of the Arctic Circle. If the eggs, however, are not thoroughly well identified, they are worthless as scientific specimens, for they cannot be separated after they have once left the nest.

Family CHARADRIIDÆ. Genus STREPSILAS.
Sub-family *SCOLOPACINÆ*.

TURNSTONE.

STREPSILAS INTERPRES (*Linnæus*).

(British: Possibly breeds; common spring and autumn coasting migrant; few, winter.)

Single Brooded. Laying season, June.

BREEDING AREA: Northern Nearctic and Palæarctic regions. The Turnstone probably breeds as far north as land extends in both hemispheres. In Europe it breeds in Iceland, Scandinavia, Denmark, and on some of the Baltic Islands, but its southern breeding limits elsewhere are very imperfectly determined. It is said to breed on Lord Howe's Island, off the coast of New South Wales, and there is ground for believing that it may do so on the Azores and the Canaries.

BREEDING HABITS: The Turnstone reaches its more southern breeding grounds in May, but not until June those in the far north. On passage it is gregarious, and even during summer is more or less social, several pairs frequently nesting in the same vicinity. The breeding grounds of this species are close to the sea, rocky coasts and islands over which tufts of coarse grass and a few bushes are scattered, being preferred. The Turnstone pairs annually, probably when the nesting grounds are reached, and its habits during this period are very similar to those of allied species, the bird uttering a trill in midair. The nest, generally placed under the shelter of a grass tuft, a plant, or a small bush, is merely a hollow lined with a few scraps of vegetable refuse. Of the actions of this bird at the nest little has been recorded, but alluring antics are said to be indulged in.

RANGE OF EGG COLOURATION AND MEASUREMENT: The eggs of the Turnstone are four in number. They are broadly pyriform in shape, smooth in texture, and vary from pale olive-green to pale buff in ground colour, blotched, spotted, and clouded with olive-brown, and dark reddish-brown, and with underlying markings of lilac-gray. On some eggs a series of net-like streaks of dark brown occur on the large end. A frequent type has the large blotches running obliquely across the surface. As usual, most of the large bold markings are on the major half of the egg; a type is not unfrequent in which the spotting is blurred or poorly defined. Average measurement, 1·6 inch in length, by 1·1 inch in breadth. Incubation is performed by both sexes, but the duration of the period is unknown.

DIAGNOSTIC CHARACTERS: Their form and size, colouration, and character of the markings are sufficiently distinct to prevent the eggs of the Turnstone from being confused with those of allied birds. It is far easier to determine these eggs by comparison with actual specimens than by any written description.

Family CHARADRIIDÆ. Genus TRINGA.
Sub-family SCOLOPACINÆ.

BONAPARTE'S SANDPIPER.

TRINGA FUSCICOLLIS, *Vieillot.*

(British : Rare abnormal autumn migrant.)

Single Brooded.   Laying season, June.

BREEDING AREA : Northern Nearctic region. Bonaparte's Sandpiper breeds throughout the Arctic regions of America, from Greenland in the east to the Mackenzie River in the west.

BREEDING HABITS : Bonaparte's Sandpiper is a migrant, and reaches its breeding grounds at the end of May, or early in June. It is gregarious and social in winter and whilst on passage, but nests in scattered pairs. The breeding haunts of this Sandpiper, so far as they are known, are situated close to the sea, on the barren grounds, or on the banks of rivers in its vicinity. Of its pairing habits nothing has been recorded. Mac Farlane met with the nest of Bonaparte's Sandpiper in this region, and describes it as merely a hollow in the ground, lined with a few dead leaves. The bird's actions at the nest appear never to have been described.

RANGE OF EGG COLOURATION AND MEASUREMENT : The eggs of Bonaparte's Sandpiper are four in number. They are pyriform in shape, smooth in texture, and vary in ground colour from olive to grayish-buff, blotched and spotted with dark reddish-brown and pale brown, and with underlying markings of gray. As usual the major half of the egg is the most heavily marked, many of the blotches becoming confluent. Average measurement, 1·25 inch in length, by ·9 inch in breadth. The duration of the period of incubation is unknown, as is also which sex performs the duty.

DIAGNOSTIC CHARACTERS: The eggs of this bird require careful identification. With the small amount of material at my command, I do not feel justified in attempting to give any distinctive characters.

Family CHARADRIIDÆ. Genus TRINGA.
Sub-family *SCOLOPACINÆ*.

PURPLE SANDPIPER.

TRINGA MARITIMA, *Gmelin*.

(British: Possibly breeds. Fairly common autumn migrant.)

Single Brooded. Laying season, May and June.

BREEDING AREA: Northern Nearctic and Palæarctic regions. The Purple Sandpiper breeds in Iceland, the Faroes, Norway, Spitzbergen, and Nova Zembla, on the Taimyr peninsula, and probably the islands off the North Siberian coasts as well as on the coasts of Bering Strait, and across Arctic America to Greenland.

BREEDING HABITS: The Purple Sandpiper migrates no further than it is compelled, and seldom wanders much south of those coasts that are free from ice during winter. It arrives at its breeding grounds in May or June, according to latitude or local conditions. The favourite nesting haunts of this Sandpiper are rarely situated far from the sea, either in the immediate neighbourhood of the beach, amongst broken ground covered with scanty herbage, or in marshy districts at the summit of adjoining hills. In the Faroes it breeds on the fells, commencing to do so before the snow has melted from the sheltered hollows and the tops of the hills. It seems probable that this species pairs for life, as there is evidence to show that it returns annually to one spot to breed. The nest is a mere hollow, lined with a few bits of dry vegetable refuse, such as moss or grass. The bird is a close sitter, sometimes remaining brooding over the eggs until nearly trodden upon, and then suddenly rising and feigning lameness.

RANGE OF EGG COLOURATION AND MEASUREMENT: The eggs of the Purple Sandpiper are four in number. They are pyriform in shape, smooth in texture, and vary in ground colour from pale olive to buffish-brown, spotted, blotched, mottled, and streaked with dark blackish-brown and reddish-brown, and with underlying markings of pale brown and lilac-gray. The eggs of this Sandpiper are very beautifully marked. One type has the markings large and bold, and scattered obliquely round the major half of the egg, the smaller markings being on the minor half; another has them large and small, evenly distributed over most of the surface; another has spots, streaks, and dark scratches intermingled with the ordinary blotches. On most the underlying markings are very conspicuous, and add much to the beauty of the egg. Average measurement, 1·5 inch in length, by 1·05 inch in breadth. Incubation is performed by both sexes, but the duration of the period is unknown.

DIAGNOSTIC CHARACTERS: The eggs of the Purple Sandpiper cannot always be distinguished from those of the Common and Jack Snipes, but the ground colour is generally much more olive. They require careful identification in many localities, where the breeding areas of these species impinge.

Family CHARADRIIDÆ.   Genus TRINGA.
Sub-family SCOLOPACINÆ.

BROAD-BILLED SANDPIPER.

TRINGA PLATYRHYNCHA, *Temminck.*

(British : Rare abnormal spring and autumn migrant.)

Single Brooded. Laying season, latter half of May, and in June.

BREEDING AREA : Northern Palæarctic region. Although the Broad-billed Sandpiper has been met with in various parts of Europe and Asia during summer, from the Atlantic to the Pacific, its breeding grounds are practically untraced. It has only been met with nesting on the Scandinavian fells as far south as lat. 60°, and in Finland. It may possibly breed in the Baikal area, and on the Stanavoi mountains in Eastern Siberia.

BREEDING HABITS : The Broad-billed Sandpiper reaches such of its breeding grounds as are known about the last week in May. It is perhaps more of a social species than a gregarious one, but it appears to breed in small if scattered colonies. It pairs annually, and, as usual, much of its courtship is carried on in the air, the bird careering about like a Snipe, and uttering a rapid note. Richard Dann met with this bird breeding in small colonies in the grassy morasses and swamps at the head of the Bothnian Gulf, and in the swamps of the Dovre-fjeld, three thousand feet above sea-level. Just after its arrival it was very wild and wary, feeding on the banks of the pools and lakes. As the breeding season approached, it became more skulking, creeping through the long grass, and when flushed, dropping again almost at once. He describes the nest as being like that of a Snipe, and made in a tuft of grass. Wolley also was fortunate enough to meet with this species. He says that its favourite nesting places were soft open spots in the marshes, where the ground was clothed with bog-moss and sedge, and the nests were often placed on tufts of grass just above the water. He describes the nest as a rounded hollow, lined with a little dry grass. The sitting bird was not only observed to run from the nest, but to fly from it, and when the eggs were much incubated, to become very tame and confiding. Nests of this Sandpiper, found by Mr. Mitchell on the Dovre-fjeld, were in open parts of the marshes, and more elaborately made than usual, the hollow being deeper and more carefully lined. He also remarked the exceedingly interesting fact that the lining in each nest resembled the colour of the eggs it contained, the darker varieties being laid on withered leaves of the willow, the paler ones on dry grass. What a pity it is other collectors do not observe more of these curious facts and minor details, especially with regard to the rarer birds, whose nidification is so little known!

RANGE OF EGG COLOURATION AND MEASUREMENT: The eggs of the Broad-billed Sandpiper are four in number. They are pyriform in shape, smooth in texture, and buffish-white in ground colour, densely mottled and spotted with rich chocolate-brown and paler brown, and with underlying markings of gray. Various distinct types occur. One has the spotting so dense and close as to hide nearly all trace of the pale ground colour; another is more thinly spotted with a larger amount of ground colour visible between the spots; another has most of the markings on the major half of the egg, many of them confluent, and the pale gray underlying spots on the minor half are very distinct and large. Average measurement, 1·3 inch in length, by ·9 inch in breadth. Incubation is performed by both sexes, but the duration of the period is unknown.

DIAGNOSTIC CHARACTERS: The eggs of the Broad-billed Sandpiper cannot readily be confused with those of any other Palæarctic species, the small size of the markings, their density and colour, being sufficient for easy identification.

Family CHARADRIIDÆ. Genus TRINGA.
Sub-family *SCOLOPACINÆ*.

AMERICAN PECTORAL SANDPIPER.

TRINGA ACUMINATA PECTORALIS (*Say*).

(British : Rare abnormal spring and autumn migrant.)

Single Brooded. Laying season, June.

BREEDING AREA: Northern Nearctic region. The American Pectoral Sandpiper breeds on the barren grounds above the limits of forest growth in Arctic America, from Alaska in the west to Davis Strait in the east.

BREEDING HABITS: The American Pectoral Sandpiper passes along the coasts and river valleys of the United States on its way north in spring, and reaches its breeding grounds towards the end of May. It migrates in flocks, but these disperse at the summer haunts after the birds have paired. Of the nesting habits of this Pectoral Sandpiper absolutely nothing was known until the eggs were obtained near Point Barrow, in Alaska, at the end of June 1883. The nest is said always to be made amongst the grass in a dry part of the tundra, but has not yet been minutely described. Of the habits of the bird at the nest nothing has yet been recorded.

RANGE OF EGG COLOURATION AND MEASUREMENT: The eggs of the American Pectoral Sandpiper are four in number. They are pyriform in shape, smooth in texture, and vary in ground colour from pale olive-brown to pale buff, spotted and blotched with rich reddish-brown, and with underlying markings of gray. As usual most of the largest markings are on the major half of the egg, where they are often confluent; the pale underlying markings are also conspicuous and well defined. Average measurement, 1·5 inch in length, by 1·1 inch in breadth. The duration of the period of incubation is unknown, as is also which sex performs the task.

DIAGNOSTIC CHARACTERS: The eggs of this Sandpiper closely resemble those of the American Stint in colour, but are much larger in size. They cannot readily be confused with those of any other species breeding in this area.

Family CHARADRIIDÆ.   Genus TRINGA.
Sub-family SCOLOPACINÆ.

LITTLE STINT.

TRINGA MINUTA, *Leisler*.

(British: Fairly common spring and autumn coasting migrant.)

Single Brooded. Laying season, latter half of June, and early in July.

BREEDING AREA: North-western Palæarctic region. The Little Stint breeds somewhat locally on the tundras above the limits of forest growth from Northern Scandinavia eastwards to the Taimyr peninsula. It has been found nesting at Kistrand, in Northern Norway, on the

Kola peninsula in Lapland, the delta of the Petchora in Russia, may probably do so on the Waigats and Nova Zembla, and is known to breed on the Yalmal peninsula, in the valley of the Yenesay, and on the Taimyr peninsula.

BREEDING HABITS: The Little Stint arrives at its breeding grounds in June, with the departure of winter, the exact date of its appearance varying locally owing to the state of the season. It migrates in flocks, but these disperse at the nesting places, although the bird is certainly a social one all through the summer, and several nests may frequently be found quite close together. Of the pairing habits of this bird nothing is known. This is not the fault of the two British naturalists who discovered its breeding grounds in the Petchora Valley in 1875, but the misfortune of their arriving at them too late, when the birds had already paired. The first naturalist to discover the nest and eggs of the Little Stint was Middendorff, who obtained them on the Taimyr peninsula in Northern Siberia, nearly fifty years ago. In 1875 Messrs. Seebohm and Harvie Brown found the breeding grounds of this Stint in Northern Russia, at the delta of the Petchora. In 1876 Finsch got a nest on the Yalmal peninsula; Hencke states that he has taken the nest near Archangel; Mr. Rae got another nest on the Kola peninsula; whilst the Swedish naturalist, Collett, found eggs near Kistrand in the Porsanger Fjord in Northern Norway. Mr. Seebohm has also obtained eggs of this Stint in the valley of the Yenesay in 1877. At the mouth of the Petchora the breeding grounds of the Little Stint were situated on a comparatively dry and gently sloping part of the tundra, close to the inland sea, at the mouth of the great river. Here the tundra was thickly studded with tussocks of grass, and the swampy ground was almost concealed by cotton grass. These grass tufts were covered with green moss, and smaller patches of reindeer-moss, the whole almost hidden with a thick growth of cloud-berry and carices, dwarf shrubs, and sundry Arctic flowers. Some of the nests were found on more sandy ground full of small pools, and covered with short grass and plants. The nest of the Little Stint, which Mr. Seebohm brought home, and which I have examined, was merely a slight hollow lined with a few dead leaves of the cloud-berry and other scraps of vegetable refuse. The female alone appears to frequent the nest, and when this is approached she makes little demonstration, and is remarkably quiet. Her tameness is often most extraordinary. At one nest she approached within eighteen inches, and when a hand was stretched towards her she quietly retreated a few feet; but the moment the nest was left she commenced to flutter along the ground as if wounded.

RANGE OF EGG COLOURATION AND MEASUREMENT: The eggs of the Little Stint are four in number. They are pyriform in shape, smooth in texture, and vary in ground colour from pale greenish-gray to pale brown, spotted and blotched with rich reddish-brown, and with underlying markings of paler brown and gray. Occasionally a few dark streaks occur on the larger end of the egg. As usual, most of the larger blotches are on the major half of the egg, and more or less confluent. Average measurement, 1·1 inch in length, by ·8 inch in breadth. Incubation is apparently performed by the female only, but the duration of the period is unknown.

DIAGNOSTIC CHARACTERS: The eggs of the Little Stint require careful identification in all localities east of the Urals, owing to the presence of allied species or races in Siberia, whose breeding areas are by no means clearly known. From European localities they are readily distinguished by their size and colour from those of all other Sandpipers, with the one exception of Temminck's Stint. As a rule, the eggs of the Little Stint are yellower than those of the latter bird, partaking more of the character of those of the Dunlin. It is easier to separate the eggs of a mixed series than to point out the differences in words.

Family CHARADRIIDÆ.   Genus TRINGA.
Sub-family *SCOLOPACINÆ*.

AMERICAN STINT.

TRINGA SUBMINUTA MINUTILLA, *Vieillot*.

(British : Very rare abnormal autumn migrant.)

Single Brooded. Laying season, latter half of June, and early in July.

BREEDING AREA: Northern Nearctic region. The American form of Middendorff's Stint breeds in the Arctic regions of America, from Alaska to Labrador and Newfoundland, and as far south as Nova Scotia.

BREEDING HABITS: The American Stint begins to arrive in the Southern States from Central and South America in April, slowly travels on to New England early in May, reaches North Carolina towards the end of the month, and appears on the Arctic "barren" grounds early in June, when they are "barren" grounds no longer, but teem with animal and vegetable life. This Stint is gregarious in winter and on migration, but separates into more or less scattered pairs for the summer. The favourite breeding haunts are the marshy moors near the sea, sometimes near the coast, more frequently a short distance inland, on the margins of the lakes and pools. Of the pairing habits of this bird I find nothing recorded, but they doubtless do not differ from those of allied species. The nest, which is merely a slight hollow lined with a little withered grass and dead leaves, is often made under the shelter of a small bush. The female is described as being very tame and trustful at the nest, but sometimes tries to allure an intruder away from its vicinity by feigning lameness.

RANGE OF EGG COLOURATION AND MEASUREMENT: The eggs of the American Stint are four in number. They are pyriform in shape, smooth in texture, and vary from pale buff to pale olive in ground colour, spotted and blotched with reddish-brown and dark brown, and with underlying markings of paler brown and gray. They appear, so far as is known, to cover exactly the same range of colour variation as those of the Little Stint and Temminck's Stint. Average measurement, 1·0 inch in length, by ·8 inch in breadth. Incubation is performed chiefly by the female, but the duration of the period is unknown.

DIAGNOSTIC CHARACTERS: The eggs of this Stint closely resemble those of the allied species, but are a little smaller than those of Temminck's Stint. The locality, if it can be thoroughly relied upon, should be sufficient to identify the eggs of the American Stint.

Family CHARADRIIDÆ. Genus TRINGA.
Sub-family SCOLOPACINÆ.

TEMMINCK'S STINT.

TRINGA TEMMINCKI, *Leisler*.

(British : Rare spring and autumn coasting migrant.)

Single Brooded. Laying season, June.

BREEDING AREA : Northern Palæarctic region. Temminck's Stint breeds on the tundras above the limits of forest growth from Scandinavia across Arctic Europe and Siberia to the Tchuski Land on the Pacific coast. In the west in Europe it breeds as far south as lat. 65° on the White Sea and Bothnian Gulf, but ten degrees lower in the far east on the shore of the Okhotsk Sea. The evidence of its breeding on the lofty mountains of Southern Siberia is not conclusive.

BREEDING HABITS : Temminck's Stint reaches its breeding grounds in Northern Europe at the end of May, and about a week later those in Northern Asia. On migration it is gregarious, but in the breeding season nests in isolated pairs, many however living within a small area of suitable ground. This Stint pairs soon after its arrival in the north. During this operation it frequently perches in small trees, or stands on a post or fence, vibrating its wings and trilling lustily. This trill, however, is generally uttered whilst the bird is wheeling round and round, or hovering and floating in the air, although it is sometimes heard as the bird runs along the ground with uplifted wings. The favourite breeding haunts of this bird are the marshy parts of the tundras, especially such where long grass and scattered dwarf willows occur near to open water. The nest is invariably made upon the ground, and is merely a hollow amongst the sedge, rushes, or grass, scantily lined with dry grass and withered leaves. When disturbed at its nesting place, this Stint becomes very demonstrative and noisy, often betraying the locality of the nest by careering wildly about above it. When it finds that its secret is actually known it becomes much quieter, unnaturally tame and confiding, and endeavours to draw all attention upon itself.

RANGE OF EGG COLOURATION AND MEASUREMENT: The eggs of Temminck's Stint are four in number. They are pyriform in shape, smooth in texture, and vary from pale buff to pale olive of various shades in ground colour, spotted and blotched with reddish-brown and dark brown, and with underlying markings of paler brown and gray. The large markings, as usual, are most numerous on the major half of the egg, sometimes forming a semi-confluent zone or irregular cap, but the smaller spots are pretty evenly distributed over most of the surface. Occasionally a few dark streaks occur on the larger end of the egg. Average measurement, 1·1 inch in length, by ·85 inch in breadth. Incubation appears to be performed by the male, but the duration of the period is unknown.

DIAGNOSTIC CHARACTERS: There is no reliable character by which the eggs of Temminck's Stint can be distinguished from those of the Little Stint and several other allied species. Careful identification at the nest is absolutely necessary.

Family CHARADRIIDÆ. Genus TRINGA.
Sub-family SCOLOPACINÆ.

SANDERLING.

TRINGA ARENARIA, *Linnæus*.

(British : Common autumn and coasting migrant.)

Single Brooded. Laying season, latter half of June.

BREEDING AREA: Northern Nearctic and Palæarctic regions. The Sanderling has only been met with nesting in the Old World in Iceland, but has been observed in summer on the Golaievskai Islands in the Petchora Gulf, on the Waigats and Nova Zembla, in the delta of the Yenesay, on the Taimyr peninsula, and the Liakoff Islands. In the New World it probably breeds in Alaska, and has been found nesting on the Anderson River (lat. 68°), on the Parry Islands (lat. 78°), in Grinnell Land (lat. 82½°), and in Greenland, on the west coast near Smith South (lat. 79°), and Godthaab (lat. 63°); on the east coast, Sabine Island (lat. 74½°).

BREEDING HABITS: The Sanderling reaches its far-away Arctic haunts late in May, or early in June, as soon as open water and bare ground can be found. It is gregarious during winter, and migrates north in flocks, but these ultimately separate into pairs, and scatter over the breeding grounds for the summer. Of the pairing habits of the Sanderling nothing has been recorded, fortunate observers woefully neglecting their opportunities, as usual. We have little information respecting the habits of this Sandpiper during the breeding season. MacFarlane was probably the first naturalist to take the eggs, killing a female from the nest in North-west America on the 29th of June, 1863. The breeding haunts of this bird are the barren grounds and tundras near and the beaches of the Arctic Ocean. Mac-Farlane's nest was discovered on the tundra about ten miles west of Franklin Bay, and was merely a hollow scantily lined with dry grass and leaves. Thirteen years afterwards, almost to the very day (24th June), Capt. Feilden found another nest close to Cape Union in Grinnell Land. This nest was made on a ridge of gravel several hundred feet above sea-level, and was a slight hollow in the centre of a bent-down willow plant, lined with a few dead leaves and withered catkins.

RANGE OF EGG COLOURATION AND MEASUREMENT: The eggs of the Sanderling are four in number. They are somewhat pyriform in shape, smooth in texture, and buffish-olive in ground colour, densely mottled, and spotted with pale olive-brown, and with underlying markings of ink-gray. Two distinct types occur, so far as our knowledge extends at present. One of these has the markings thickly congregated on the major half of the egg; the other is more uniformly marked over the entire surface. Average measurement, 1·4 inch in length, by 1·0 inch in breadth. Incubation is performed by both sexes, but the duration of the period is unknown.

DIAGNOSTIC CHARACTERS: The peculiar character of the markings (small and mottled), combined with the shape and size, readily distinguish the eggs of the Sanderling from those of allied species.

Family CHARADRIIDÆ. Genus TRINGA.
Sub-family SCOLOPACINÆ.

BUFF-BREASTED SANDPIPER.

TRINGA RUFESCENS, *Vieillot.*

(British : Very rare abnormal autumn migrant.)

Single Brooded. Laying season, June and early July.

BREEDING AREA : Northern Nearctic region, and possibly extreme north-eastern Palæarctic region. The Buff-breasted Sandpiper breeds in the Arctic regions of America, from Alaska probably to Baffin Bay, and south to about lat. 53°. It is just possible that this species may cross Bering Strait, and breed on the Siberian coasts.

BREEDING HABITS : The Buff-breasted Sandpiper travels to and from its breeding grounds by inland rather than coast routes, crossing the prairies, and arriving in the Arctic regions early in June. On migration it is certainly gregarious, but the flocks apparently disperse in scattered pairs over the breeding grounds, although many nests may be found within a small area of suitable country. Of the habits of this bird in the nesting season but little has been recorded, not from want of opportunities on the part of naturalists, but from lack of observation. MacFarlane, who found this bird breeding in large numbers on the barren ground between the Horton River and the shores of the Arctic Ocean, states that the nest is always made upon the ground, and resembles that of the American Golden Plover. Other observers describe it as a slight hollow scantily lined with dry grass. Mr. Murdoch also met with this Sandpiper breeding at Point Barrow in Alaska, and says that it frequented the dry portions of the barren grounds, the nest being a mere hollow lined with a little moss. The parent birds are said to be very tame at the nest, only flying for a little distance when flushed from the eggs.

RANGE OF EGG COLOURATION AND MEASUREMENT: The eggs of the Buff-breasted Sandpiper are four in number. They are pyriform in shape, smooth in texture, and vary in ground colour from pale buff to deep buff, sometimes tinged with olive, blotched and spotted with rich reddish-brown, and with numerous underlying markings of ink-gray. As usual, most of the larger blotches are on the major half of the egg, and often confluent, and types occur in which the markings are diagonally distributed. Occasionally a few dark-brown streaks occur on the larger end of the egg; the underlying markings are very distinctly defined. Average measurement, 1·45 inch in length, by 1·0 inch in breadth. The duration of the period of incubation is unknown, as is also which sex performs the task.

DIAGNOSTIC CHARACTERS: The eggs of this bird require very careful identification, as they closely resemble those of other allied species breeding within the same area; and from which I am at present unable to give any thoroughly reliable character to distinguish them.

Family CHARADRIIDÆ. Genus SCOLOPAX.
Sub-family *SCOLOPACINÆ.*

GREAT SNIPE.

SCOLOPAX MAJOR, *Gmelin.*

(British : Rare abnormal spring and autumn migrant.)

Single Brooded. Laying season, late in May and in June.

BREEDING AREA: Western Palæarctic region. The Great Snipe breeds in Holland, Denmark, Northern Germany, Poland, throughout Scandinavia, and in Russia from as far south as lat. $50°$, north to the coast. In Western Siberia it appears to breed as far north as lat. $67\frac{1}{2}°$ in the valley of the Obb, but only to $66\frac{1}{2}°$ in the valley of the Yenesay, which is probably its eastern limit.

BREEDING HABITS: The Great Snipe arrives at its Scandinavian breeding grounds during the latter half of May, but in the Arctic regions further east in the first half of June. Whilst on migration and during the pairing season the Great Snipe is somewhat gregarious. When mating, the males appear often to collect into parties, and go through various strange antics in the air and on the ground. Mr. Seebohm relates how he has watched them "stretch out their necks, throw back the head almost upside down, and open and shut their beaks rapidly, uttering a curious noise like that produced by running the finger along the edge of a comb." Sometimes these notes were uttered just after the bird had taken a short flight, or spread its wings and tail. As many as six birds were counted in the air together, during this singular tournament, in another locality. The favourite breeding grounds of this Snipe are in swamps, especially those in which bare patches of mud, peat, or sand occur; and the marshy margins of rivers and lakes, where tall rank grass-tufts, sedges, and other aquatic vegetation occur. As soon as pairing is completed, the birds scatter over such districts to breed. The nest is either made amongst the long coarse grass, or in the centre of a tuft of rush or sedge, and is a mere hollow lined with dry grass or a little moss. The bird sits closely, and usually flies right away when flushed from the nest.

RANGE OF EGG COLOURATION AND MEASUREMENT: The eggs of the Great Snipe are four in number. They are pyriform, smooth in texture, and vary in ground colour from olive and grayish-buff to brownish-buff, spotted and blotched with rich dark brown and pale brown, and with underlying markings of lilac-brown and gray. The eggs are exceedingly handsome ones. Most of the blotches are obliquely distributed, and on many eggs net-like masses of dark-brown streaks occur on the larger end. Most of the larger blotches, some of them confluent, are on the major half of the egg, and the underlying markings are generally large and conspicuous. Average measurement, 1·8 inch in length, by 1·25 inch in breadth. Incubation, performed chiefly if not entirely by the female, lasts seventeen or eighteen days.

DIAGNOSTIC CHARACTERS: Their large size and handsome and oblique character of the markings, readily distinguish the eggs of this Snipe from those of every other British species.

Family CHARADRIIDÆ. Genus SCOLOPAX.
Sub-family *SCOLOPACINÆ.*

JACK SNIPE.

SCOLOPAX GALLINULA, *Linnæus.*

(British : Common autumn and winter migrant.)

Single Brooded. Laying season, June.

BREEDING AREA: Northern Palæarctic region. The Jack Snipe breeds on the tundras above the limits of forest growth in Arctic Europe and Asia ; on the swamps of the Dovre-fjeld and Lapland, and in Western Russia to as far south as St. Petersburg. In Asia it apparently breeds as high as lat. 70°, and probably as low, in some districts, as lat. 60°.

BREEDING HABITS : The Jack Snipe reaches its more southerly breeding places in May, but does not arrive in the extreme northern ones until June. It is a solitary bird, and breeds in isolated pairs. The Jack Snipe apparently mates in spring, and after arriving in its breeding haunts. It is doubtful whether the male drums like the Common Snipe at this season, although much of its courtship takes place in the air. I should say, judging from the description given by Wolley and Naumann, that the note is similar to that uttered by the Great Snipe under sexual excitement, and made in the same way by the bird rapidly opening and shutting its beak. Our information respecting the nidification of the Jack Snipe was principally obtained by Wolley in Lapland. He found nests of this bird, after a most persevering and patient search, placed in dry spots amongst the sedge and grass close to the borders of the more open swamps. They were mere hollows lined with a little dry grass, *equisetum*, and dead withered leaves of the dwarf birch. The bird is a close sitter, its plumage closely resembling surrounding tints, and remains brooding over the eggs until almost trodden upon. Wolley was allowed to approach one nest within six inches before the bird rose.

RANGE OF EGG COLOURATION AND MEASUREMENT: The eggs of the Jack Snipe are four in number, and very large for the size of the bird, a clutch weighing nearly as much as the hen herself. They are pyriform in shape, and vary from buff to olive in ground colour, blotched and spotted with rich reddish- or blackish-brown, and with underlying markings of pale brown and gray. These eggs are very handsome ones. Most of the larger markings are on the major half of the egg, where they are often confluent. A few streaks of blackish-brown also frequently occur on this part, whilst the pale underlying markings, from their number, size, and distinctness, add, by contrast, to the beauty of the egg. Average measurement, 1·5 inch in length, by 1·0 inch in breadth. Incubation appears to be performed by the female, but the duration of the period is unknown.

DIAGNOSTIC CHARACTERS: The eggs of the Jack Snipe require the most careful identification, as no character can be given by which they can be distinguished from those of the Common Snipe, the Buff-breasted Sandpiper, the Purple Sandpiper, and some other allied species. If not thoroughly well authenticated, they are simply valueless as scientific specimens.

Family STERCORARIIDÆ. Genus STERCORARIUS.

BUFFON'S SKUA.

STERCORARIUS BUFFONI (*Boie*).

(British : Rather rare nomadic autumn and spring migrant.)

Single Brooded. Laying season, early in July.

BREEDING AREA : Northern Nearctic and Palæarctic regions. Buffon's Skua breeds on the tundras and barren grounds above the limits of forest growth in both hemispheres, and in a similar climate on the treeless moors of the Dovre-fjeld, as far south as lat. 62½°. It breeds in Lapland, probably Spitzbergen, on Nova Zembla, and thence across the tundras of Europe and Asia eastwards to Bering Strait. In the New World it breeds on the barren grounds of the Arctic regions of America, from Alaska to Greenland.

BREEDING HABITS : Buffon's Skua reaches its Arctic breeding grounds during the first half of June. At all seasons this Skua is more or less gregarious : it breeds in colonies which are, however, scattered, and cover a wide area of ground. Of the pairing habits of this bird nothing appears to have been observed or recorded. The breeding haunts of this Skua are the most barren portions of the tundras at no great distance from the Arctic Ocean, where the grass, moss, and lichens that clothe the ground grow in scattered patches with strips of bare sand or clay between them. The nest is merely a shallow hollow amongst the moss or grass, scantily lined with dead leaves of the dwarf birch or the cranberry, and a little dry grass or moss. The birds are very bold and pugnacious at their breeding grounds, resenting any intrusion, and according to MacFarlane the female occasionally shams lameness when disturbed from her eggs.

RANGE OF EGG COLOURATION AND MEASUREMENT: The eggs of Buffon's Skua are usually two in number, very rarely one or three. They are sometimes pyriform, sometimes rotund in shape, and vary in ground colour from olive to brown of various shades, spotted, speckled, and streaked with dark brown, and with a few faint underlying markings of grayish-brown. They cover the same range of colour variation as those of Richardson's Skua. Average measurement, 2·0 inches in length, by 1·4 inch in breadth, Incubation is performed by the female, and is said to last about a month.

DIAGNOSTIC CHARACTERS: The eggs of Buffon's Skua resemble those of Richardson's Skua, but are much smaller, more constantly green in ground colour, and the markings are, as a rule, a trifle more streaky. The breeding area of Buffon's Skua is also much more northerly.

Family STERCORARIIDÆ.　　　　Genus STERCORARIUS.

POMATORHINE SKUA.

STERCORARIUS POMATORHINUS (*Temminck*).

(British: Fairly common autumn and spring coasting migrant; few, winter.)

Single Brooded. Laying season, July.

BREEDING AREA: Northern Nearctic and Palæarctic regions. The Pomatorhine Skua is not known to breed anywhere on continental Europe, but may probably do so on Franz-Josef Land, Spitzbergen, and Nova Zembla. Eastwards it has been met with nesting on the Yalmal and Taimyr peninsulas, and in the north-east of Siberia, whilst it very probably breeds on the Liakoff Islands off the coast of that country in the Arctic Ocean. On the American continent it breeds in Alaska, and probably on the barren grounds across to Greenland, in which latter country it is known to do so in Egedes Land on the east coast in about lat. 67°.

BREEDING HABITS: The Pomatorhine Skua is a bird of somewhat restricted migration, wintering, as a rule, as far north as it can with safety. It arrives at its Arctic breeding grounds during the first half of June. But little has been recorded of the habits of this bird during the breeding season. Von Middendorff found colonies of this Skua nesting near the Taimyr Lake on the peninsula of that name in Northern Siberia; and Finsch met with it breeding on the Yalmal peninsula in the same country. It appears to breed in colonies. Of its pairing habits I find nothing recorded. The nest is merely a hollow in the moss- and lichen-covered ground of the tundra. The behaviour of the birds when their breeding grounds are invaded by man is not known to differ from that of allied species.

RANGE OF EGG COLOURATION AND MEASUREMENT: The eggs of the Pomatorhine Skua are two in number, sometimes but one. They vary in ground colour from pale olive to dark olive, with a buffish tinge, spotted and speckled with dull reddish-brown, and with underlying markings similar in character of grayish-brown. The spots are not very large, nor very distinctly defined, are irregular in shape, and usually most numerous and largest on the major half of the egg. Average measurement, 2·4 inches in length, by 1·7 inch in breadth. Incubation is apparently performed by the female, but the duration of the period is unknown.

DIAGNOSTIC CHARACTERS: The eggs of the Pomatorhine Skua are larger than those of Richardson's Skua, but in some cases otherwise closely resemble them in colour. They also bear a close resemblance to certain varieties of those of the Common Gull, but the locality should be sufficient to determine them, as that bird does not breed within the same area.

Family LARIDÆ.
Sub-family *LARINÆ*.

Genus PAGOPHILA.

IVORY GULL.

PAGOPHILA EBURNEA (*Phipps*).

(British : Rare nomadic autumn and winter migrant.)

Single Brooded. Laying season, June and early July.

BREEDING AREA : Extreme Northern Nearctic and Palæarctic regions. The Ivory Gull is confined practically to the land in the North Polar basin during the breeding season, and nests as far south as Franz-Josef Land, Spitzbergen, Nova Zembla, Herald Island, the Parry Islands, and Grinnell Land.

BREEDING HABITS : The Ivory Gull does not wander much south of open water during winter, and retires to the Polar regions as early as it can find food. Our information respecting the nidification of this Gull is only of the most meagre description. Of its pairing habits absolutely nothing has been recorded. It is apparently gregarious during the breeding season. Its favourite nesting places are lofty precipices above the sea, but in some localities where such cannot be had, it resorts to the sea-beach, or to a low island in the Polar Sea. It appears sometimes to nest in company with allied birds, such as

Kittiwakes and Glaucous Gulls. The nests found by Dr. Malmgren on the cliffs of Spitzbergen are described as shallow hollows in the soil on the ledges, lined with dry grass, moss, weeds, and a few feathers. Those found by M'Clintock on the Parry Island group were made upon the ground entirely of moss, but one of them contained a few feathers and a little white down.

RANGE OF EGG COLOURATION AND MEASUREMENT: The eggs of the Ivory Gull are never more than two in number, and very frequently appear only to be one. They vary in ground colour from olive-brown to buffish-brown, spotted with dark brown and pale brown, and with underlying markings of lilac-gray. In the distribution of the markings two types occur, one in which they are pretty evenly distributed over the entire surface of the egg, and the other in which they are mostly on the major half, where they sometimes form an irregular zone. Average measurement, 2·5 inches in length, by 1·7 inch in breadth. The duration of the period of incubation is unknown, as is also which sex performs the duty. Professor Collett has described a fine series of the eggs of this Gull in the *Ibis* (1888, p. 440).

DIAGNOSTIC CHARACTERS: The eggs of this Gull somewhat closely resemble those of the Common Gull in colour, but are larger in size. The locality of the eggs should be sufficient to determine their identity, as the Common Gull does not breed within the same area. The eggs of the Kittiwake are also much smaller.

Family LARIDÆ.   Genus LARUS.
Sub-family LARINÆ.

ICELAND GULL.

LARUS LEUCOPTERUS, *Faber.*

(British : Rare nomadic autumn and winter migrant.)
Single Brooded.   Laying season, June and early July.

BREEDING AREA: Northern Nearctic region. The Iceland Gull is only known to breed in Greenland, and probably across Arctic America to Alaska and the Aleutian Islands.

BREEDING HABITS: The Iceland Gull is another species that wanders but a short distance south of open water during winter, and retires to its breeding haunts as soon as the adjoining seas are free from ice. This Gull breeds more or less in colonies, and its favourite nesting places are either the lofty cliffs above the Arctic Ocean or the sandy beaches at the mouths of rivers that fall into that sea ; in some districts small islands appear to be preferred. Of the pairing habits of this Gull I find nothing recorded. The only nests of this species that appear to have been described are those found by Dall on the banks of the Yukon in Alaska. These were merely shallow hollows in the sand.

RANGE OF EGG COLOURATION AND MEASUREMENT: The eggs of the Iceland Gull are two or three in number. They vary in ground colour from pale grayish-buff to dark buff and olive, blotched and spotted with dark brown and pale brown, and with underlying markings of lilac-gray. The markings are seldom large, and either uniformly distributed over the surface of the egg or mostly confined to the major half, where they not unfrequently form a zone. The gray underlying markings are large, and both numerous and well-defined. Average measurement, 2·75 inches in length, by 1·8 inch in breadth. The duration of the period of incubation is unknown, as is also which sex performs the duty.

DIAGNOSTIC CHARACTERS: The eggs of the Iceland Gull cannot always be distinguished from those of the Lesser Black-backed Gull and the Herring Gull, but as a rule they are not nearly so blotched as those of the former species, nor ever so dark in ground colour as those of the latter. The locality of the eggs should settle their identity at once, as the breeding areas of these two species do not impinge upon that of the Iceland Gull.

Family LARIDÆ.   Genus LARUS.
Sub-Family *LARINÆ*.

GLAUCOUS GULL.

LARUS GLAUCUS, *Fabricius*.

(British : Irregular nomadic winter migrant.)

Single Brooded. Laying season, June.

BREEDING AREA : Northern Nearctic and Palæarctic regions. The Glaucous Gull breeds in Iceland, Spitzbergen, Nova Zembla, at Vardö in North-eastern Norway, and along the shores of the Arctic Ocean across Europe, Asia, and America to Greenland, where it ranges as far north as lat. 82°.

BREEDING HABITS : The Glaucous Gull is a nomadic migrant, wandering little from the vicinity of its breeding area during winter, and retiring north as soon as the season permits. It is a gregarious bird, and breeds in colonies of varying size. This species probably pairs for life, and returns to the same nesting places each season, although further information is much to be desired. This Gull breeds on cliffs in some localities, on the ground in others. In some districts low flat islands of sand in the deltas of rivers are resorted to. When placed on the ledges of precipices it is a somewhat massive structure composed of dead seaweed and drift, and lined with dry grass, but when on the ground it is little more than a shallow cavity lined with a few bits of dry grass. Nests found by Messrs. Seebohm and Harvie-Brown on a sand-bank in the delta of the Petchora, are described by them as heaps of sand hollowed slightly at the apex, and lined with a few tufts of coarse sea-weed. MacFarlane states that they were mere depressions in the beach. This Gull is bold and pugnacious at the nesting grounds, rising in flocks from the eggs, and circling above the intruder, or swooping past him uttering noisy cries.

RANGE OF EGG COLOURATION AND MEASUREMENT: The eggs of the Glaucous Gull are two or three in number. They are somewhat rough in texture, and vary in ground colour from pale olive-brown of various shades to pale bluish-green, spotted and blotched with dark brown and with underlying markings of brownish-gray. As a rule, the spots are not very large, and fairly distributed over most of the surface of the egg, but less frequently the markings are moderately large blotches mostly on the major half, where they sometimes form an irregular zone. I have examined eggs from Vardö, which are said to belong to the Glaucous Gull, pale red in ground colour, spotted with darker red and with underlying markings of lilac-gray. Average measurement, 3·0 inches in length, by 2·0 inches in breadth.

Incubation lasts a month, but whether both sexes or which performs the duty is apparently unknown.

DIAGNOSTIC CHARACTERS: Unfortunately I know of no character by which the eggs of the Glaucous Gull can be distinguished from those of the Herring Gull or from small examples of those of the Great Black-backed Gull. They require careful identification.

Family LARIDÆ.
Sub-family *LARINÆ*.

Genus LARUS.

GREAT BLACK-HEADED GULL.

LARUS ICHTHYAËTUS, *Pallas*.

(British : Very rare abnormal spring migrant.)

Number of Broods unknown. Laying season, May and June.

BREEDING AREA: South-central Palæarctic region, and North-eastern Ethiopian region. The Great Black-headed Gull breeds in Egypt, Cyprus, Palestine, on the Seal Islands in the Caspian Sea, also on the low-lying coasts of that sea itself, northwards to the lakes of Turkestan, South-western Siberia, Western Mongolia, and Thibet.

BREEDING HABITS: But little is known respecting the nidification of this Gull. It is a migrant in the more northern portions of its range, reaching them for breeding purposes in May or early in June. It breeds on the low-lying coasts and islands as well as on the banks of inland lakes. Of its pairing habits nothing is known, nor can I find that its nest has been accurately described by any competent naturalist.

RANGE OF EGG COLOURATION AND MEASUREMENT: The eggs of the Great Black-headed Gull are three in number. They are pale buff or drab in ground colour, blotched, spotted, and streaked with various shades of reddish-brown, sometimes approaching black in intensity, and with underlying markings of paler brown and gray. Average measurement, 3·0 inches in length, by 2·0 inches in breadth. The duration of the period of incubation is unknown, as is also which sex performs the duty.

DIAGNOSTIC CHARACTERS: The pale ground colour, small and dark markings, and large size of the eggs of this Gull prevent them from being readily confused with those of any other species breeding in the same area.

Family LARIDÆ.   Genus LARUS.
Sub-family *LARINÆ*.

MEDITERRANEAN BLACK-HEADED GULL.

LARUS MELANOCEPHALUS, *Natterer*.

(British : Very rare abnormal winter migrant.)

Single Brooded. Laying season, May.

BREEDING AREA: South-western Palæarctic region. The breeding grounds of the Mediterranean Black-headed Gull are most imperfectly known. This bird may breed on the Atlantic coasts of Southern France, and certainly does so on those of South-western Spain— in the delta of the Guadalquivir. It is also presumed to breed in various parts of the basin of the Mediterranean, but the only known locality appears to be Smyrna. It is also said to breed on the coasts of the Black Sea.

BREEDING HABITS: I find but little information recorded of the habits of this bird during the nesting season. They apparently differ very slightly from those of the nearly allied Black-headed Gull, which breeds so commonly in many parts of the British Islands. It breeds in colonies. Its favourite nesting places appear to be marshes, flat islands in the deltas of rivers, and low-lying coasts. The nest of this bird is described by Mr. Dresser as being made of twigs, placed on the ground, generally under or near a low scrubby bush.

RANGE OF EGG COLOURATION AND MEASUREMENT: The eggs of the Mediterranean Black-headed Gull are two or three in number, and closely resemble those of the Black-headed Gull in colour, but are generally more inclined to buff than olive. Average measurement, 2·0 inches in length, by 1·4 inch in breadth. The duration of the period of incubation is undetermined, but is probably a little over three weeks, as in the commoner species. Probably both male and female perform the duty.

DIAGNOSTIC CHARACTERS: I know of no more constant character by which the eggs of this Gull can be distinguished from those of allied species than that of smaller size. I have examined but few eggs of this Gull, and therefore am not prepared to say whether any other and more reliable character exists.

Family LARIDÆ.
Sub-family *LARINÆ*.

Genus LARUS.

BONAPARTE'S GULL.

LARUS PHILADELPHIA, *Ord*.

(British : Very rare abnormal winter and spring migrant.)

Single Brooded. Laying season, June, and first half of July.

BREEDING AREA : Northern Nearctic region. Bonaparte's Gull breeds in the sub-arctic regions of America, from Alaska in the west to Labrador in the east. It appears not to breed anywhere much above the Arctic Circle, and as far south as Manitoba, presuming that the species was correctly identified by Mr. Raine, who visited colonies on Crescent Lake and elsewhere in this region.

BREEDING HABITS : Bonaparte's Gull is a migratory bird, and reaches its breeding grounds in May. It is very gregarious, and breeds in colonies of varying size, the birds apparently returning to certain places every year. There can be little doubt that this bird pairs for life. The breeding grounds of this Gull are near lakes and pools, especially such as are close to trees and bushes, and on the margins of prairie lakes and sloughs. The most interesting feature in the nidification of Bonaparte's Gull is the fact of the birds nesting in bushes and trees. This, however, is not a universal practice, and in districts where such sites cannot be got, they breed in marshes on the ground, as for instance on the margins of the lakes on the prairies, or on low sandy islands in those lakes. In the latter situations the nest of this Gull is merely a shallow cavity in the sand, lined with a little dry grass. The nest in a tree or a bush is a much more substantial structure. When in a tree it is usually made on a flat horizontal branch at some distance from the trunk, and from fifteen to twenty feet from the ground, and when in a bush is seldom less than four feet from the ground. Several nests are frequently made in the same tree. The nest is made of sticks and twigs, and lined with dry grass, moss, or bits of dry reed. MacFarlane describes one of the nests (out of thirty-seven taken) as "composed of down and velvety leaves, held together by some stringy turf." When disturbed at their nests, the birds rise and fly to and fro in alarm, uttering anxious cries.

RANGE OF EGG COLOURATION AND MEASUREMENT: The eggs of Bonaparte's Gull are two or three in number. They vary considerably in ground colour, from rich brown through every shade to yellowish-buff on the one hand, and to olive-brown and olive-green on the other, spotted with dark brown, and with underlying markings of pale brown and brownish-gray. The spots are seldom large, and pretty evenly distributed over the entire surface of the egg, but as a rule the larger ones are on the major half, where also the markings generally are most numerous, and sometimes form an irregular zone. Average measurement, 2·0 inches in length, by 1·4 inch in breadth. Incubation is performed by both sexes, and lasts from twenty-one to twenty-four days.

DIAGNOSTIC CHARACTERS: The eggs of Bonaparte's Gull closely resemble those of the Black-headed Gull, but are smaller. The locality of course will distinguish them in doubtful cases. They also resemble those of Franklin's Gull, but are smaller.

Family LARIDÆ. Genus LARUS.
Sub-family LARINÆ.

LITTLE GULL.

LARUS MINUTUS, *Pallas*.

(British : Irregular nomadic autumn and winter migrant.)

Single Brooded. Laying season, June.

BREEDING AREA: Northern Palæarctic region. The Little Gull formerly bred in Göttland, also not far from Dantzic, and, according to Herr E. Hartert, may still do so in some parts of North-eastern Germany. It breeds in Esthonia, more commonly on the lakes of Ladoga and Onega in Western Russia, probably in the vicinity of Archangel, and abundantly in the morasses of the Ural. Eastwards it breeds across Southern Siberia to the Stanavoi Mountains and the shores of the Ochotsk Sea.

BREEDING HABITS: The Little Gull is a rather late migrant, not arriving at its breeding grounds until they are free from ice in May, or early in June, according to locality. The favourite breeding haunts of this Gull are inland swamps and lakes, especially such as contain islands of semi-floating weed and aquatic plants. It is gregarious, and breeds in colonies of varying size. As it returns each season to certain localities, there seems little doubt that this Gull pairs for life. In many of the Baltic colonies it breeds in company with the Common Tern. The nests, which are often made close together, are placed amongst the coarse grass, sedge, reeds, and other aquatic vegetation on the margin of the lake, or upon the masses of floating weed and *débris* at some distance from the shore. The nests are well-made and somewhat bulky structures, composed externally of dead reeds, sedge, and grass, and lined with finer sedge and dry grass. When disturbed the birds rise in a fluttering throng, and fly round and round above their nests uttering their peculiar short, screeching note, and behave generally like the Black-headed Gull under similar circumstances.

RANGE OF EGG COLOURATION AND MEASUREMENT: The eggs of the Little Gull are generally three, but in rare instances four in number. They vary in ground colour from pale buffish-brown to olive-brown and olive-green, spotted and more rarely blotched and streaked with dark brown, and with underlying markings of pale gray. As a rule, the spots are not very large, and the blotches, formed by one or two spots becoming confluent, are not numerous. Average measurement, 1·7 inch in length, by 1·2 inch in breadth. Incubation is performed by both sexes, but the duration of the period is apparently unknown.

DIAGNOSTIC CHARACTERS: Unless the eggs of the Little Gull are thoroughly well authenticated they are absolutely worthless. I know of no character which will distinguish the buff or brown varieties from those of the Common Tern. The nest, however, is very different; it may also be remarked that the eggs of the Tern rarely, if ever, show any trace of olive or green in the ground colour.

Family LARIDÆ. Genus XEMA.
Sub-family LARINÆ.

SABINE'S GULL.
XEMA SABINII (*J. Sabine*).
(British : Rare nomadic autumn migrant.)
Single Brooded. Laying season, June and July, according to locality.

BREEDING AREA : Northern Nearctic and Palæarctic regions. Sabine's Gull is either very locally distributed during the breeding season, or its nesting places are most imperfectly known, notwithstanding its circumpolar summer area. It appears to breed in Spitzbergen, and was observed nesting by Middendorff on the Taimyr peninsula. It may also breed on the Siberian coasts of Bering Strait, whence it is said to nest in suitable localities across Arctic America from Alaska to Baffin Bay, including Western Greenland, where this species was first discovered by Sir Edward Sabine, who found it breeding in about lat. $75\frac{1}{2}°$. MacFarlane records that a large number of nests were found on the shores of Franklin Bay.

BREEDING HABITS : Sabine's Gull wanders no further south in winter than absolutely compelled by stress of weather, and returns to its usual Arctic haunts as soon as they become free from ice in June. Its favourite breeding grounds are the marshy tundras and barren grounds close to the Polar Sea, especially on some peninsula, or even on an island in a lake or river near the coast, or occasionally at some distance from it. But little is known of the habits of this Gull during the breeding season. It is gregarious and nests in colonies, often in company with the Arctic Tern. Von Middendorff describes the nest as a mere hollow in the moss, lined with a little dry grass; Sabine found the eggs on the bare ground; whilst Richardson describes it as a hollow in the short mossy turf. When disturbed at the nest, the birds rise in an anxious fluttering crowd, uttering an incessant note and flying to and fro until left in peace.

RANGE OF EGG COLOURATION AND MEASUREMENT: The eggs of Sabine's Gull, so far as is known, are two in number. They vary in ground colour from pale brown to dark brown, occasionally tinged with olive, spotted with darker brown, and with very indistinct underlying markings of grayish-brown. On some eggs a few short streaks occur on the larger end. Average measurement, 1·7 inch in length, by 1·25 inch in breadth. The duration of the period of incubation is unknown, as is also the sex performing the duty.

DIAGNOSTIC CHARACTERS: The eggs of this Gull somewhat resemble the more sparingly marked eggs of the Arctic Tern in colour, but are distinguished by their larger size and darker ground colour.

Family LARIDÆ.   Genus ANOUS.
Sub-family STERNINÆ.

NODDY TERN.

ANOUS STOLIDUS (*Linnæus*).

(British: Very rare abnormal autumn migrant.)

Double Brooded (?). Laying season, May and June (Northern Hemisphere), September to January (Southern Hemisphere).

BREEDING AREA: Oceanic tropical zone. The Noddy Tern breeds on the various islands and reefs that stud the tropic seas right round the world. Amongst the principal breeding stations may be mentioned those on the West Indies, off the coasts of Florida (Bahamas, Tortugas), and Central America, various islands in Polynesia and off the coasts of Australia, the Laccadives and other isles in the Indian Ocean, various islands in the Red Sea, and Tristan d'Acunha, St. Helena, and Ascension in the tropical Atlantic Ocean.

BREEDING HABITS: According to latitude the Noddy Tern visits its breeding stations in May or September to breed. It is a very gregarious species, and some of its colonies are very extensive. Of the pairing habits of this bird I find nothing recorded, but probably the union endures for life, as the same colonies are frequented year after year. The site for the nest varies a good deal according to locality. In some districts the nest is placed on a mangrove or in the crown of a cocoa-nut palm; in others on the ledges beneath overhanging precipices; in others, but more rarely, on level patches of sand or on grass-covered slopes. It is described as often being a large structure, made of dry grass, sea-weed, and twigs rudely heaped together, with a shallow cavity at the top for the eggs.

RANGE OF EGG COLOURATION AND MEASUREMENT: The eggs of the Noddy Tern are probably three in number. There seems to be some doubt about this, several naturalists asserting that the full clutch consists of one egg only; but as the eggs of this Tern are gathered systematically for food in every well-known breeding station, it is very probable the eggs are taken as they are laid. (Conf. *Ibis*, 1891, p. 146.) They are rough and chalky in texture, and vary from reddish-white to pale buff in ground colour, spotted and blotched with reddish-brown, and with underlying markings of pale brown. The markings are not very numerous nor large, and are usually pretty evenly distributed over the surface of the egg. Average measurement, 2·0 inches in length, by 1·4 inch in breadth. The duration of the period of incubation is unknown, as is also the sex which performs the task.

DIAGNOSTIC CHARACTERS: The eggs of this Tern cannot readily be confused with those of any other "British" species, but I am not prepared to say whether they can be distinguished from those of allied birds.

Family LARIDÆ. Genus STERNA.
Sub-family STERNINÆ.

SOOTY TERN.

STERNA FULIGINOSA, *Gmelin*.

(British : Very rare abnormal spring (?) and autumn migrant.)

Double Brooded(?). Laying season, September to January (Southern Hemisphere), April and May (Northern Hemisphere).

BREEDING AREA: Oceanic tropical zone. The Sooty Tern, like the Noddy, breeds on the various islands that dot the tropic seas right round the world. Amongst its principal breeding places may be mentioned the islands off the coast of Florida, the West Indies, the island of Ascension, St. Helena, Madagascar, Rodriguez, Mauritius, the Chagos Archipelago, the Laccadives, the islands in the Red Sea and Persian Gulf, Ceylon, many parts of the Malay Archipelago, the coasts of Australia, various islands throughout Polynesia, and in Lower California.

BREEDING HABITS: The Sooty Tern visits its various breeding places for purposes of reproduction in May and September, according to locality and latitude. It is a gregarious species, some of its colonies being very extensive, but of its pairing habits nothing appears to have been recorded. Probably it mates for life, as it uses the same localities year by year. One of the most remarkable breeding places of this Tern is situated on the island of Ascension, and consists of two small colonies and one very large one. Here the eggs are so thick upon the ground that it is a difficult matter to walk amongst them without treading on them. The nest is merely a hollow in the ground. I have seen photographs of this curious and interesting colony, which resembles the usual breeding places of sea-birds, say at the Farne Islands or Scoulton Mere; that is to say, the ground covered with nests (the Terns all sitting head to wind), and the air clouded with birds. Vast numbers of eggs are taken in Ascension for food—three thousand being picked up in a single morning—and the laying season is abnormally prolonged, as at the Farne Islands, as the poor deluded birds are induced to replace their losses.

RANGE OF EGG COLOURATION AND MEASUREMENT: The eggs of the Sooty Tern are normally three in number. (Conf. *Ibis*, 1891, p. 145.) They are smooth in texture, and vary in ground colour from white, through cream to pale buff, spotted with reddish-brown, and with underlying markings of pale brown and gray. The spots are not very large, but are irregular in shape, and somewhat evenly distributed over most of the surface of the egg, occasionally forming an irregular zone round the major half. Average measurement, 2·0 inches in length, by 1·5 inch in breadth. The duration of the period of incubation is unknown, as is also the sex which performs the duty.

DIAGNOSTIC CHARACTERS: The eggs of the Sooty Tern somewhat closely resemble certain types of those of the Sandwich Tern, but the spots are never so dark in colour. From eggs of the Noddy Tern they are at once distinguished by their smooth shell.

Family LARIDÆ. Genus STERNA.
Sub-family STERNINÆ.

CASPIAN TERN.

STERNA CASPIA, *Pallas*.

(British : Rare abnormal spring and autumn migrant.)

Single Brooded. Laying season, May and June.

BREEDING AREA: Temperate and tropical zones. The Caspian Tern (with the exception of South America, and where it may possibly yet be found to breed, if it is not entirely replaced by the nearly allied *Sterna maxima*) breeds on the coasts, and by some inland waters in various districts, right round the world. Amongst its breeding places may be mentioned the island of Sylt, off the west coast of Schleswig, various low coasts and isles in the basin of the Baltic, possibly on the south coast of Holland, on the eastern coasts of Spain, on various islets near Sardinia, and elsewhere in the basin of the Mediterranean, on the lagoons of the Danube, and elsewhere on the shores of the Black and Caspian Seas. It also breeds in the deltas of the Nile and the Zambesi, on various islands in the Persian Gulf, and by the salt lakes of Turkestan. It is known to breed in Ceylon, in Australia, and New Zealand. In the New World it ranges about as far north as in Western Europe, namely, to Labrador and Alaska, and southwards to California and Mexico.

BREEDING HABITS: The Caspian Tern returns to its accustomed breeding places early or late in May according to locality. This Tern breeds in societies, but many of its European colonies are sadly reduced in numbers owing to incessant plundering of the nests. The favourite breeding grounds are on the shores of lagoons, on low sandy islands, and on the flat, sandy shores of inland salt-lakes. Doubtless this Tern pairs for life, returning as it does with great regularity to certain breeding places. The nest is of the scantiest description, a mere hollow in the sand, sometimes, but not always, with a few bits of sea-weed or dry grass round the rim. When disturbed at their breeding place the birds rise and hover in the air above their nests, becoming very clamorous, and boldly swooping down round the head of the intruder.

RANGE OF EGG COLOURATION AND MEASUREMENT: The eggs of the Caspian Tern are two or three in number. They vary in ground colour from creamy-white to buffish-brown, spotted with dark brown, and with underlying markings of gray. The surface markings are not very large, and generally pretty evenly distributed, but sometimes most numerous on the major half of the egg, and forming an irregular zone. Occasionally a few streaky spots occur amongst the others, and the underlying markings are often large and generally conspicuous. Average measurement, 2·55 inches in length, by 1·7 inch in breadth. Incubation, according to Meyer, lasts about twenty days, but whether both parents assist in the duty is not known.

DIAGNOSTIC CHARACTERS: The eggs of the Caspian Tern are distinguished from those of all other Terns by their large size, with the possible exception of those of *S. maxima*.

Family LARIDÆ. Genus STERNA.
Sub-family STERNINÆ.

GULL-BILLED TERN.

STERNA ANGLICA, *Montagu*.

(British : Rare abnormal spring and autumn migrant.)

Single Brooded. Laying season, latter end of April, and in May and June.

BREEDING AREA : Southern Palæarctic region, North-western Oriental region, North-eastern Ethiopian region, and Eastern Nearctic region. So far as I can determine, the breeding area of the Gull-billed Tern is absolutely discontinuous, the isolated individuals being those that breed in the Nearctic region. The Gull-billed Tern breeds on the island of Sylt off the west coast of Schleswig, and in Denmark. There appear to be no other northern breeding places in Europe, and the bird's next stations are in the south of Spain and in the delta of the Rhone in Southern France. Thence we trace its colonies through Greece, Asia Minor, and the basins of the Black and Caspian Seas. South of the Mediterranean it breeds locally in Northern Africa from Morocco to the shores of the Red Sea. Eastwards in Asia it breeds by the lakes of Persia, Turkestan, Cashmere, and Southern Mongolia. In the New World it is generally distributed along the Atlantic coasts of North America from about lat. 40° south to Texas, Mexico, and the West Indies.

BREEDING HABITS : The Gull-billed Tern reaches its southern breeding grounds in Europe in April, but is a month later in the more northerly ones. It is gregarious like its congeners, breeding in colonies, and returning to certain spots year by year, so that it probably pairs for life. Its favourite breeding grounds are the shores of lagoons, and low islands in deltas, the shores of lakes, sandy islands, and banks. The nest is merely a slight hollow scraped out in the sand, and sometimes lined with a few bits of dry sea-weed and grass, but in some cases the eggs are deposited in some natural hollow, without any further provision or alteration. When disturbed at the colony the birds rise in a fluttering crowd, become very noisy, and fly to and fro in great anxiety until left to themselves again.

RANGE OF EGG COLOURATION AND MEASUREMENT: The eggs of the Gull-billed Tern are two or three in number. They vary in ground colour from creamy-white to buffish-brown, spotted with various shades of brown, and with underlying markings of gray. The spots are not very large—varying from the size of buck-shot downwards—and usually pretty evenly distributed over the surface of the egg, but sometimes most numerous on the major half. The gray underlying markings are similar in character, and very conspicuous. Average measurement, 2·0 inches in length, by 1·4 inch in breadth. Incubation is performed by both sexes, but the duration of the period is unknown.

DIAGNOSTIC CHARACTERS: The eggs of the Gull-billed Tern are not easily confused with those of any other Palæarctic species. Small eggs similar in colour might be confused with those of the Sandwich Tern, but are always duller in colour.

Family LARIDÆ. Genus HYDROCHELIDON.
Sub-family *STERNINÆ*.

WHISKERED TERN.

HYDROCHELIDON HYBRIDA (*Pallas*).

(British : Very rare abnormal spring and autumn migrant.)

Single Brooded. Laying season, May to July, according to locality.

BREEDING AREA: Tropical and sub-tropical regions of the Eastern Hemisphere. The Whiskered Tern breeds in Spain, very sparingly in the south of France (Rhone delta), in the valley of the Danube (and it is said sparingly at Lublin in Poland), in Southern Russia, Turkey, Greece, Asia Minor, Palestine, Armenia, Turkestan, Cashmere, Northern India, and Mongolia. It also breeds in North-eastern Australia, and may probably do so in the Philippines and the Malay Archipelago. South of the Mediterranean it breeds in Northern Africa from Morocco to Egypt, and may possibly do so in the Transvaal and elsewhere in the southern half of the African continent.

BREEDING HABITS: The Whiskered Tern visits its European breeding places late in April, but does not breed in India until June or July. It is gregarious, and breeds in colonies like its allies, probably pairing for life, as it returns yearly to certain haunts. The favourite breeding grounds of this Tern are marshes, especially those situated in or near deltas, or near lakes and large rivers. The nest is a somewhat bulky structure, composed of rushes, sedges, and dry grass, the latter forming the lining. In Algeria, Canon Tristram observed this species breeding in the old nests of the Eared Grebe. Sometimes the nest is floating—as observed by Anderson in

India—but is more generally amongst the coarse vegetation of the marsh. The actions of this Tern when disturbed at the nest are similar to those of allied species.

RANGE OF EGG COLOURATION AND MEASUREMENT: The eggs of the Whiskered Tern are two or three in number. They vary in ground colour from pale buff to pale grayish-green, blotched and spotted with various shades of reddish-brown and blackish-brown, and with underlying markings of pale brown and gray. The markings are neither very numerous nor very large, and as a rule pretty evenly distributed over the surface of the egg. On some eggs the markings are elongated into short streaks intermixed with faint scratches and scrawls. Average measurement, 1·55 inch in length, by 1·15 inch in breadth. Incubation is apparently performed by both sexes, but the duration of the period is undetermined.

DIAGNOSTIC CHARACTERS: The eggs of this Tern are readily separated from those of the two following species by their larger size, smaller markings, and paler and greener ground colour. Some eggs of the Arctic Tern approach them in colour, but are generally more heavily marked and larger.

Family LARIDÆ. Genus HYDROCHELIDON.
Sub-family *STERNINÆ*.

BLACK TERN.

HYDROCHELIDON NIGRA (*Linnæus*).

(British : Formerly bred ; regular spring and autumn coasting migrant.)

Single Brooded. Laying season, latter end of May, and in June.

BREEDING AREA : South-western Palæarctic Region. The typical form of the Black Tern breeds as far north as the province of Esthonia south of the Gulf of Finland, thence southwards through the Baltic Provinces, Prussia, Southern Scandinavia, Denmark, Holland, France, and the Spanish peninsula, and eastwards throughout Central and Southern Europe in all suitable districts, to the Caspian. In Asia it breeds in South-western Siberia and Turkestan, to as far east as the Altai Mountains. South of the Mediterranean it breeds throughout North Africa with the exception of Egypt. The Nearctic form of this species is known as *Hydrochelidon nigra surinamensis*.

BREEDING HABITS: The Black Tern reaches its accustomed breeding places in May. It is a gregarious bird, nesting in colonies of varying size, and sometimes in company with the following species. There can be little doubt that it pairs for life. The favourite breeding haunts are marshes and weed-grown shallow lakes and pools, either near the sea on lagoons, or in inland localities. The nests are made either amongst the reeds or *equisetums* at some distance from the bank in shallow water, or amongst the hassocks of sedge and other coarse vegetation in the marshes. They are large and bulky structures—heaps of half-rotten reeds, sedges, and other marsh-loving plants, lined with finer and drier material, such as sedge, leaves of the rushes and reeds, and dry grass. The behaviour of this Tern when its breeding grounds are invaded is similar to that of allied species, the birds rising in noisy crowds and fluttering to and fro in an anxious manner, until the cause of their alarm withdraws.

RANGE OF EGG COLOURATION AND MEASUREMENT: The eggs of the Black Tern are three in number. They vary in ground colour from clay colour to olive-brown or olive-green, passing through almost every shade of buff, blotched and spotted with reddish-brown and dark blackish-brown, and with underlying markings of pale brown and gray. As a rule the blotches are most numerous on the major half of the egg, and are often confluent, but types occur in which the markings are smaller and more evenly distributed over the entire surface. Average measurement, 1·4 inch in length by 1·0 inch in breadth. Incubation is performed by both sexes, but the duration of the period is apparently undetermined.

DIAGNOSTIC CHARACTERS: The eggs of the Black Tern cannot be distinguished from those of the White-winged Black Tern, and unless identified at the nest are of no scientific value. It might be remarked, however, that the Black Tern breeds in the Spanish peninsula, but the White-winged Black Tern is not known to do so. Locality in such cases is sufficient for identification.

Family LARIDÆ. Genus HYDROCHELIDON.
Sub-family STERNINÆ.

WHITE-WINGED BLACK TERN.

HYDROCHELIDON LEUCOPTERA (*Meisner and Schinz*).

(British : Rare abnormal spring and autumn migrant.)

Single Brooded. Laying season, latter end of May, and in June.

BREEDING AREA: Southern Palæarctic region, and possibly extreme north-eastern Ethiopian region. The White-winged Black Tern breeds in Poland, and in various suitable localities in Central and Southern Europe, including Sicily, Northern Italy, the delta of the Rhone, Hungary, and Southern Russia. South of the Mediterranean it is said to breed in Algeria and Lower Egypt. Eastward it breeds in suitable districts in the Caucasus, and may do so in Turkestan. It also breeds throughout the extreme southern districts of Siberia to the Amoor, southwards to Northern China and Mongolia.

BREEDING HABITS: The White-winged Black Tern visits its wonted breeding places early in May. It is an inland species, and, like the other members of this genus, frequents marshes rather than sands. This Tern is also gregarious, breeding in colonies by itself in many localities, but in others socially resorting with its ally the Black Tern. It probably pairs for life. The favourite breeding grounds are shallow pools full of reeds and rushes, and the swampy shores of lakes and pools. The nest is either placed amongst the reeds, sedges, and other coarse vegetation near the shore, or on more or less floating masses of weed and aquatic plants at some distance from the bank in shallow water. It is a somewhat large, strong, and bulky structure, made of half-rotten sedges, reeds, and *equisetums*, and lined with bits of dry reeds, leaves of the rushes, or coarse grass. When disturbed at its breeding places this Tern becomes very anxious and noisy, rising in fluttering crowds, and keeping in the air until left in peace.

RANGE OF EGG COLOURATION AND MEASUREMENT: The eggs of the White-winged Black Tern are three in number, but four are said to be laid in rare instances. I have never seen a clutch of this latter number, but have taken several clutches of four belonging to the Lesser Tern. They cover precisely the same range of colour variation as those of the preceding species, rendering a detailed description unnecessary. Average measurement, 1·4 inch in length by 1·0 inch in breadth. Incubation is performed by both sexes, but the duration of the period is apparently unknown.

DIAGNOSTIC CHARACTERS: The eggs of the White-winged Black Tern cannot be distinguished from those of the Black Tern, and therefore, to be of any scientific value, must be carefully identified at the nest.

Family ALCIDÆ.   Genus URIA.

BRUNNICH'S GUILLEMOT.

URIA TROILE BRUNNICHI, *Sabine*.

(British: Possibly breeds: Very rare nomadic autumn migrant.)

Single Brooded generally. Laying season, June.

BREEDING AREA: North Atlantic and Arctic Ocean basins. Brunnich's Guillemot breeds probably from the Liakoff Islands westwards to Nova Zembla, Spitzbergen, Franz-Josef Land, Grimsey Island north of Iceland, and Greenland, north of about lat. 64°.

BREEDING HABITS: Brunnich's Guillemot is a nomadic migrant, and wanders little south of open water during winter, retiring north again as soon as the Polar seas are free from ice. It is a gregarious bird, and breeds in colonies, some of these nesting places being tenanted by vast numbers of birds. Probably it pairs for life, as season after season the same haunts are frequented. Like the Common Guillemot the present sub-species breeds on the ledges of the cliffs that overhang the sea, gathering at the old accustomed places in May and June. Its habits are not known to differ in any important respect from those of the southern race. It makes no nest, and lays its eggs on the ledges and in the hollows on the shelves of the beetling cliffs. When disturbed at the colony the birds leave the cliffs in streams, and resort to the water below, making no demonstration of alarm, or showing any resentment at the intrusion of their haunts.

RANGE OF EGG COLOURATION AND MEASUREMENT: Brunnich's Guillemot lays only one egg, but if this be taken, it is usually replaced several times in succession. It covers precisely the same range of colour variation as that of the Common Guillemot, fully described in *Nests and Eggs of British Birds*, p. 311. Average measurement, 3·2 inches in length, by 2·2 inches in breadth. Incubation is performed by both sexes, but the duration of the period is undetermined. Possibly this may be slightly different from the period occupied by the Common Guillemot, when we bear in mind the much colder climate in which the eggs are hatched.

DIAGNOSTIC CHARACTERS: The eggs of Brunnich's Guillemot cannot be distinguished from those of the Common Guillemot in colour, but are generally broader and blunter in shape. The locality is also of some service in correctly identifying them.

Family ALCIDÆ. Genus MERGULUS.

LITTLE AUK.

MERGULUS ALLE (*Linnæus*).

(British : Irregular nomadic autumn and winter migrant.)

Single Brooded. Laying season, June.

BREEDING AREA : North Atlantic and Arctic Ocean basins. The Little Auk breeds in colonies of varying size in Nova Zembla, Franz-Josef Land, Spitzbergen, Grimsey Island to the north of Iceland, and the east and west coasts of Greenland, from north of about lat. 68°, to about lat. 79°.

BREEDING HABITS: The Little Auk is another species which wanders no further south in winter from its Polar haunts than the necessities of life compel it. It is gregarious and usually breeds in vast colonies, some of them probably containing hundreds of thousands of pairs. It spends the greater part of its life at sea, but in May resorts to the land to breed. There can be little doubt that this Auk pairs for life, and yearly returns to one place to rear its young. Its breeding grounds are not always situated by the sea, but are sometimes at a considerable distance from the coast, and are more sloping rock-covered banks than precipitous cliffs. A favourite site is on the sloping ground below a range of cliffs, where the bank is covered with stones and broken rocks that have from time to time crumbled from the precipices towering above. Dr. Hayes met with a very large colony of this bird on the Greenland coast of Smith Sound, situated on the slopes of both sides of a rocky valley which was crowned with lofty cliffs. In its breeding habits this bird very closely resembles the Puffin. It makes no nest, however, and lays its eggs under large stones and rock fragments, and in hollows and holes under the *débris*. When disturbed at the nesting place the birds leave their eggs in swarms, and fly towards the sea. It is a noisy bird, and is even said to utter its notes whilst sitting under the rocks and stones.

RANGE OF EGG COLOURATION AND MEASUREMENT: The Little Auk lays only one egg each season; whether this is renewed if taken appears not to have been observed. It is pale greenish-blue, sometimes very indistinctly streaked with yellowish-brown. Average measurement, 1·9 inch in length, by 1·25 inch in breadth. Incubation is performed by both sexes, but the duration of the period is unknown.

DIAGNOSTIC CHARACTERS: The egg of the Little Auk cannot readily be confused with those of any other species breeding in the same area, its size and colouration easily distinguishing it. Whether it can be separated so surely from the eggs of certain allied species breeding in the North Pacific region, I am unable to say.

Family ALCIDÆ.   Genus ALCA.

GREAT AUK.

ALCA IMPENNIS, *Linnæus*.

(British: Formerly bred. Now extinct.)

Number of Broods unknown   Laying season, June.

FORMER BREEDING AREA: North Atlantic basin. The Great Auk there is, alas, only too much reason to believe is now extinct, although the evidence of its extinction is purely negative, and the regions which it formerly inhabited are by no means completely explored. This curious flightless bird used to breed on the coasts of Newfoundland (especially on Funk Island) and Labrador, Southern Greenland, Iceland, St. Kilda, the Faroes, and possibly the coasts of Norway.

BREEDING HABITS: The habits of the Great Auk only possess an historical interest. Unfortunately the bird ceased to exist before the era dawned when the habits of birds were studied minutely and in careful detail, so that our information is of a general character only. So far as can be gathered, the Great Auk somewhat closely resembled the Razorbill in its habits, due allowance being made for its flightless condition. It was a gregarious bird, and appeared to breed in colonies, or in scattered pairs amongst its more flourishing congeners. Its breeding places were on such rocks that could be reached without the aid of wings, sloping to the water, and consequently in many places further inland than the Razorbill selected. Whether the Great Auk made a nest is not known, but the probabilities are against it.

RANGE OF EGG COLOURATION AND MEASUREMENT: The Great Auk is presumed to have laid only one egg each season, for Martin quaintly says of the bird at St. Kilda, "he flyeth not at all, lays his (*sic*) egg upon the bare rock, which, if taken away, he lays no more for that year." Only about seventy eggs of this bird are known to exist. I have examined perhaps half-a-dozen of the eggs themselves, and about a dozen carefully-painted models. They resemble the common type of Razorbill's eggs, being pyriform, rough in texture, yellowish-white or pale brown in ground colour, blotched and spotted with brown of various shades, sometimes almost black in hue, and with underlying markings of paler brown and gray. One type has the blotches large and boldly defined, mostly situated on the major half of the egg; another type is sparingly spotted, chiefly on the major half; another type is beautifully and intricately pencilled and streaked over most of the surface. Average measurement, 4·9 inches in length, by 2·8 inches in breadth. The duration of the period of incubation is unknown, as is also which sex performed the duty.

DIAGNOSTIC CHARACTERS: The size, shape, and general colouration of the eggs of the Great Auk are sufficiently distinctive to prevent confusion with those of any other known species.

Family PROCELLARIIDÆ.   Genus BULWERIA.

BULWER'S PETREL.

BULWERIA COLUMBINA (*Moquin-Tandon*).

(British : Very rare abnormal spring migrant.)

Single Brooded probably. Laying season, May and June.

BREEDING AREA: Northern Circumtropical Seas. Bulwer's Petrel is only known to breed on the Canaries, Madeira, and Porto Santo Islands, on the Sandwich Islands, and on the Volcano Islands.

BREEDING HABITS: Bulwer's Petrel is an oceanic bird, and only resorts to the land to breed. Its favourite nesting places are isolated island groups with a good ocean aspect. This bird probably pairs for life, and resorts to certain spots each season for breeding purposes. No nest is made, the egg being laid in hollows under rock fragments, or beneath large stones at the foot of the cliffs, at a distance of several feet from the open. The bird is a close sitter, and shuns the light of day.

RANGE OF EGG COLOURATION AND MEASUREMENT: Bulwer's Petrel only produces a single egg for a sitting. This is pure white. Average measurement, 1·75 inch in length, by 1·25 inch in breadth. Incubation is performed by both sexes, but the duration of the period has not been ascertained.

DIAGNOSTIC CHARACTERS: The size and absence of spotting is sufficient to distinguish the egg of Bulwer's Petrel from those of any other species breeding in the same area.

Family PROCELLARIIDÆ.   Genus OCEANITES.

WILSON' PETREL.
OCEANITES WILSONI (*Bonaparte*).

(British : Rare summer visitor during the period of the Antipodean winter.)

Single Brooded. Laying season, January and February (Southern Hemisphere).

BREEDING AREA: Southern Seas. Wilson's Petrel is only known to breed on Kerguelen Island, but probably does so on the various isolated rocks and islands that stud the Southern Ocean right round that portion of the globe.

BREEDING HABITS: The only information respecting the nidification of Wilson's Petrel is that obtained by the Rev. A. E. Eaton, the naturalist to the Transit of Venus Expedition, on Kerguelen Island. This Petrel is gregarious at its breeding grounds, and there can be little doubt that it pairs for life, being observed in pairs a month or more before the eggs were laid. The birds arrived at their breeding haunts towards the end of November, and as usual kept pretty close during the day, going out to sea at dusk to feed. The favourite sites for the colonies were rock and stone-strewn slopes on the sides and summits of high hills, or amongst the shattered rocks just above high-water mark on the beach. No nest is made, the egg being laid on the bare ground in hollows under stones or beneath masses of rock. When the colonies were visited at night, the birds were found to be noisy, and flying to and from their holes, uttering their notes even whilst on their eggs, but when alarmed they became silent, as if anxious not to betray their whereabouts.

RANGE OF EGG COLOURATION AND MEASUREMENT: Wilson's Petrel only produces a single egg for a sitting. This is white in ground colour, generally with a more or less distinct zone of dust-like specks of reddish-brown round the larger end. Average measurement, 1·3 inch in length, by ·9 inch in breadth. Incubation is performed by both sexes, but the duration of the period is undetermined.

DIAGNOSTIC CHARACTERS: The eggs of this Petrel cannot be distinguished from those of the Fork-tailed Petrel, but as the latter only breeds in the Northern Hemisphere, the locality is sufficient to separate them. Whether the eggs can be distinguished from those of other Petrels breeding in the Southern Hemisphere, lack of material prevents me from stating.

Family PROCELLARIIDÆ.   Genus PUFFINUS.

SOOTY SHEARWATER.

PUFFINUS GRISEUS (*Gmelin*).

(British : Rare summer visitor during the period of the Antipodean winter.)

Single Brooded probably. Laying season, December and January.

BREEDING AREA : Southern Seas. The Sooty Shearwater is at present only known to breed on the Chatham Islands, but there can be no doubt that many other places remain to be discovered.

BREEDING HABITS : As is unfortunately the case with so many species in the present family, but little is known of the nidification of the Sooty Shearwater. The bird spends the greater portion of the year at sea, but as the breeding season approaches, resorts to certain spots for purposes of reproduction. Of the pairing habits of this Shearwater nothing appears to have been observed, but doubtless the bird mates for life. It is gregarious and breeds in scattered colonies. The nest is placed at the end of a burrow made in the soft ground, and is merely a slight collection of twigs and dry leaves. The burrow, according to Mr. Travers, runs for some three or four feet nearly straight, and then turns for a short distance to the right or left. This Shearwater is nocturnal in its habits, and is described as very noisy at night at its breeding places. Like its congeners it is a close sitter, and is consequently undemonstrative at the nest.

RANGE OF EGG COLOURATION AND MEASUREMENT : The Sooty Shearwater produces only a single egg for a sitting. This is somewhat smooth in texture, oval in form (varying from a rotund to an oblong oval), and pure white. Some difference of opinion exists as to the correct measurements of the eggs of this Shearwater. I am disposed to accept those given by Sir Walter Buller in his *History of the Birds of New Zealand*, as correct, especially as they have been recently confirmed by Mr. H. O. Forbes, who gives a series of dimensions in the *Ibis* for 1893, p. 542. The measurements given by Saunders in his *Manual of British Birds*, and by some other writers, are too small, and must refer to the eggs of other species. Average measurement, 3·1 inches in length, by 2·0 inches in breadth. Incubation is apparently performed by both sexes, but the duration of the period is unknown.

DIAGNOSTIC CHARACTERS: Our knowledge of the eggs of the Shearwaters is too limited and too uncertain to allow of any characters being given, even if such exist, by which they may be distinguished. Size and locality are of some service, but unless the eggs are thoroughly well identified they are not of the slightest scientific value.

Family PROCELLARIIDÆ.   Genus PUFFINUS.

DUSKY SHEARWATER.

PUFFINUS OBSCURUS (*Gmelin*).

(British : Very rare abnormal spring migrant.)

Single brooded probably. Laying season, December to May, according to latitude.

BREEDING AREA: Tropic seas. The Dusky Shearwater breeds in Madeira and the Canaries, probably the

Azores, in the Barbadoes, Bahamas, and the Galapagos Islands, and possibly elsewhere in the islands of the Pacific.

BREEDING HABITS: The habits of the Dusky Shearwater are very imperfectly known, but what few have been observed are very similar to those of the Manx Shearwater. The bird is gregarious during the nesting season, congregating at various islands to breed, and probably pairing for life. It is a strictly nocturnal bird, scarcely ever being seen abroad during daylight (even infested by eyeless parasites), and as a consequence its actions are little observed. It does not appear to make any nest, but to deposit its egg in a burrow in the ground or in hollows under rocks.

RANGE OF EGG COLOURATION AND MEASUREMENT: The Dusky Shearwater only lays one egg for a sitting. It is rough and chalky in texture, oval in shape, and pure white. Average measurement, 2·1 inches in length, by 1·4 inch in breadth. Incubation is performed by both sexes, but the duration of the period is unknown.

DIAGNOSTIC CHARACTERS: The egg of the Dusky Shearwater is much smaller than that of the Manx Shearwater, but whether it can be distinguished from those of other allied species, I am not prepared to say. The eggs of no birds are so little known as those of the species in the present family. The locality is of some assistance in identifying them, but as the breeding areas of these birds are little known, even this is a doubtful guide.

Family COLYMBIDÆ.   Genus COLYMBUS.

GREAT NORTHERN DIVER.

COLYMBUS GLACIALIS, *Linnæus*.

(British : Possibly breeds ; fairly common nomadic autumn and winter migrant.)

Single Brooded. Laying season, June.

BREEDING AREA : Northern Nearctic region, extreme North-western Palæarctic region. The Great Northern Diver is not known to breed anywhere in Europe except in Iceland. It is an American species, breeding from Greenland westwards across the Nearctic region, south of the Arctic Circle to Alaska, and southwards to the Northern United States from Dakota to Michigan, New York, and Maine.

BREEDING HABITS : The Great Northern Diver reaches its breeding grounds towards the end of May, or early in June, as soon as the waters it frequents are free from ice. It is a nomadic species, and wanders no further south than the necessities of life demand. The favourite breeding haunts of this Diver are the wild and secluded northern lakes and tarns. It is not gregarious, and breeds in isolated pairs which show no social tendencies. There can be little doubt that this Diver pairs for life, and is much attached to certain breeding places. Wherever possible, an island is preferred for a nesting site. The nest varies a good deal in construction, according to the locality frequented. Where the ground is marshy, it is large and flat, and composed of a heap of half-rotten sedges, rushes, reeds, and similar aquatic vegetation, lined with drier and finer material, such as bits of broken reeds and withered grass. Where the shore of the lake is dry and bare, the nest is little more than a hollow in the sand or hard ground, with perhaps a few bits of dry grass by way of lining, and is frequently placed in a very exposed situation. Whilst one bird is sitting, the other keeps in the vicinity to give notice of any approaching danger. The bird is a light sitter, and usually slips off the nest the moment it is alarmed, taking refuge in the water, where it is joined by its mate. The note of this Diver during the breeding season is a wild unearthly scream.

RANGE OF EGG COLOURATION AND MEASUREMENT: The eggs of the Great Northern Diver are almost invariably two in number, but it is said three have been found in one nest. I consider this very doubtful. They vary in ground colour from russet-brown to olive-brown, spotted with dark blackish-brown, and with underlying markings of paler brown. The spots are not numerous, and mostly congregated on the major half of the egg, ranging from the size of a buck-shot downwards. The underlying markings are small and few. Average measurement, 3·5 inches in length, by 2·5 inches in breadth. Incubation, performed by both sexes, lasts about a month.

DIAGNOSTIC CHARACTERS: The eggs of the Great Northern Diver may be generally distinguished from those of the Red- and Black-throated Divers by their larger size, but exceptionally small eggs cannot be separated from those of the latter species. Careful identification is therefore necessary. From eggs of the White-billed Diver I am unable to give any character by which they may be distinguished, as the eggs of that bird still remain unknown to science. (*Conf.* Appendix I. p. 344.)

Family PODICIPEDIDÆ.   Genus PODICEPS.

BLACK-NECKED GREBE.

PODICEPS NIGRICOLLIS, *Brehm.*

(British : Abnormal spring and autumn migrant.)

Single Brooded. Laying season, latter half of May, and early June.

BREEDING AREA: Southern Palæarctic region, Ethiopian region. The Black-necked Grebe breeds in Prussia, and elsewhere in Northern Germany, and is said to do so in Denmark. Southwards it breeds in suitable localities throughout Central and Southern Europe and Russia, as far north as Moscow. It breeds commonly in the basin of the Mediterranean, and is said to do so in Africa as far south as Damara Land, the Transvaal, and Cape Colony. Eastwards it breeds in Southern Siberia, sparely in Turkestan, and probably in Mongolia and Manchooria.

BREEDING HABITS: The Black-necked Grebe is a migrant in the colder and more northern portions of its range, returning to its breeding haunts as soon as they are free from ice, which in Europe is towards the end of March or early in April, but in Asia some weeks later. This Grebe cannot be regarded as generally gregarious or even social; it lives in scattered pairs, and even in districts where it is abundant each pair keep much to themselves, although other nests may be in the immediate vicinity. The bird pairs for life. In the love season the note of this Grebe, usually heard on warm, still evenings, is so quickly repeated as to sound almost like a trill. The breeding haunts of the Black-necked Grebe are fresh-water lakes and slow-running rivers whose shores and banks are clothed with reeds, sedge, and other aquatic vegetation. The nest is generally a floating structure amongst the reeds or rushes, but is frequently placed on a hassock of sedge or coarse grass at some distance from the open water. The nest is moderately large and compact, and composed of half-rotten sedges, rushes, reeds, and aquatic plants massed and heaped together, the cavity containing the eggs being lined with finer and drier material, leaves of the rushes, bits of reed, and dry grass. It would appear that in some districts more gregarious instincts prevail; for Canon Tristram states that he found a densely-crowded colony of this Grebe on Lake Halloula in Algeria, the nests being so close together in some places as almost to touch each other. Some of these nests were made on foundations that reached from the bottom of water more than a yard in depth. When leaving the nest voluntarily the parent bird carefully covers the eggs with pieces of moss or wet grass to conceal them from view.

RANGE OF EGG COLOURATION AND MEASUREMENT: The eggs of the Black-necked Grebe are four or five in number. They are rough in texture, almost as much pointed at one end as the other, and yellowish-white, sometimes with obscure traces of the green interior showing through. Average measurement, 1·8 inch in length, by 1·2 inch in breadth. Incubation, performed by both sexes, lasts from twenty-one to twenty-four days.

DIAGNOSTIC CHARACTERS: The eggs of the Black-necked Grebe are readily distinguished from those of the Little Grebe by their larger size, but are absolutely indistinguishable from those of the Sclavonian Grebe. They can be separated from those of the Red-necked Grebe by their smaller size, the two measurements on any single egg never overlapping. It should be remarked that the breeding areas of the present species and the Sclavonian Grebe are to a great extent distinct, so that the locality is of some assistance in identifying the eggs.

Family PODICIPEDIDÆ. Genus PODICEPS.

SCLAVONIAN GREBE.

PODICEPS CORNUTUS (*Gmelin*).

(British : Possibly breeds : Nomadic autumn and winter migrant.)

Single Brooded. Laying season, June.

BREEDING AREA : Northern Nearctic and Palæarctic regions. The Sclavonian Grebe breeds in Iceland, the Faroes, and throughout Europe in suitable localities north of about lat. 54° up to the Arctic Circle, except in Scandinavia, where owing to the influence of the Gulf Stream its range extends beyond that limit. Eastwards it breeds in South-western Siberia, the Baikal area, Dauria, and the valley of the Amoor. On the American continent it breeds in Alaska, and throughout British North America as far north as the Arctic Circle. It may also breed in the south of Greenland.

BREEDING HABITS : The Sclavonian Grebe is a migrant, and returns to its summer haunts late in April or during May, according to locality and state of the season. I cannot find that this Grebe is gregarious during the breeding season ; it lives in more or less scattered pairs, each keeping to themselves although nesting in the vicinity of others. Its favourite breeding grounds are lakes and pools, where the sides are shallow and clothed with a luxuriant growth of reeds, sedges, and other aquatic vegetation. This Grebe also pairs for life, but whether the same nest is used each season I am unable to say. The nest, which is a large, flat structure, is usually floating amongst the reeds and other vegetation, but sometimes is built upon a tussock in very shallow water. It is merely a heap of half-rotten sedge, rush, reed, and grass, the cavity containing the eggs being lined with the finer and drier material. The parent bird covers her eggs for concealment with bits of reed or grass upon leaving them voluntarily, even before she begins to sit, or the full complement is laid. If flushed from the nest the bird usually takes refuge in the water.

RANGE OF EGG COLOURATION AND MEASUREMENT: The eggs of the Sclavonian Grebe are four or five in number, sometimes only three; and occasionally as many as six. Mr. Raine records a clutch of the latter number taken at Long Lake near Lake Winnipeg, on the 18th of June. They are rough in texture, pointed at both ends, and yellowish-white in colour—pea-green when held up to the light, and viewed through the hole where the contents have been removed. Average measurement, 1·8 inch in length, by 1·2 inch in breadth. Incubation is performed by both sexes, and lasts from twenty-one to twenty-four days.

DIAGNOSTIC CHARACTERS: The eggs of the Sclavonian Grebe are easily distinguished from those of the Little Grebe by their larger size, but are indistinguishable from those of the Black-necked Grebe. They are smaller than those of the Red-necked and Great Crested Grebes.

Family PODICIPEDIDÆ. Genus PODICEPS.

RED-NECKED GREBE.

PODICEPS RUBRICOLLIS (*Gmelin*).

(British : Nomadic autumn and winter migrant.)

Single Brooded. Laying season, May and June.

BREEDING AREA: Western Palæarctic region. The typical form of the Red-necked Grebe breeds in Scandinavia as far north as the Arctic Circle, and in Russia, from as far north as Archangel southwards to the Caspian and Black Seas, but apparently absent from the northeast of that country. It also breeds in the remainder of Europe which lies north of the Danube valley and east of the Rhine. South of the Mediterranean, it is said to breed sparingly in Morocco and Algeria. Eastwards in Asia it breeds in South-western Siberia and Turkestan. *Podiceps rubricollis major* of Temminck and Schlegel (the *Podiceps holboelli* of Reinhardt) is the East Asian and Nearctic representative race.

BREEDING HABITS: The Red-necked Grebe wanders no further from its breeding haunts during winter than the necessities of life demand. It returns north again with the break-up of the ice, reaching its summer quarters late in March, or in April and May, according to latitude and state of the season. To a certain extent this Grebe is gregarious, breeding not only in odd pairs, but in scattered colonies, and continuing social right through the summer. Its favourite breeding places are lakes and ponds with shallow margins overgrown with reeds, rushes, and other aquatic vegetation. There can be little doubt that the bird pairs for life. The nests are usually floating structures made amongst the reeds at some distance from the actual shore, sometimes being very exposed in spots where the vegetation is thin or not fully grown. They are flat, large, and bulky structures, mere heaps of rotting reeds, sedges, grasses, and rushes, with the cavity containing the eggs lined with drier and finer material. When disturbed at the nest, the sitting bird slips quietly off into the water, but before leaving the eggs she carefully conceals them by covering them with weeds.

RANGE OF EGG COLOURATION AND MEASUREMENT: The eggs of the Red-necked Grebe are three or four in number, but sometimes as many as six. Mr. Raine records a clutch of the latter number belonging to the Nearctic form of this Grebe. They are rough in texture, elliptical in shape, and yellowish-white in colour, but traces of the green interior lining frequently show upon the surface. Average measurement, 2·0 inches in length, by 1·3 inch in breadth. Incubation, performed by both sexes, lasts from twenty-one to twenty-four days.

DIAGNOSTIC CHARACTERS: The eggs of the Red-necked Grebe are smaller than those of the Great Crested Grebe, and larger than those of the remaining European species. The measurements may overlap, but never both on the same egg—a short egg is broad, a long one may be narrow.

THE NESTS AND EGGS OF

Family RALLIDÆ.   Genus CREX.
Sub-family RALLINÆ.

LITTLE CRAKE.

CREX PARVA (*Scopoli*).

(British : Possibly breeds : Rare abnormal spring and autumn migrant.)

Single Brooded.   Laying season, May.

BREEDING AREA : Western Palæarctic region. The Little Crake breeds in Holstein, and along the southern shores of the Baltic to Livonia, and across Central and Southern Russia to Astrakhan in the valley of the Volga, and to the Caucasus. It probably breeds in Denmark, and does so in Southern Germany, Austro-Hungary, Italy, Sicily, Savoy, Central and Southern France, and possibly in Spain and Greece. Eastwards it breeds in Russian Turkestan and in Afghanistan. South of the Mediterranean it breeds in Algeria, and probably elsewhere in North-west Africa.

BREEDING HABITS : The Little Crake returns to such of its haunts as a severe climate compels it to leave in winter, in April. It is not a gregarious bird, and lives in isolated pairs, which keep exclusively to themselves. It probably pairs annually. The favourite breeding haunts of this Crake are in marshes, swamps, reed-beds, and the rank, dense vegetation on the margins of lakes and pools. The nest is artfully concealed amongst the aquatic vegetation, sometimes placed a foot or more above the surface of the water on a bunch of fallen reeds, sometimes under the shelter of a tuft of sedge. It is a somewhat large and bulky structure, as is usual with nests made in aquatic sites, and is composed of bits of reeds and flags, dry grass, rush-leaves, and flower-heads, the cavity containing the eggs being shallow, and lined with rush-leaves or dry coarse grass. The bird sits very warily, slipping quietly off into the adjoining belts of reeds or into the water with little or no demonstration of anxiety.

RANGE OF EGG COLOURATION AND MEASUREMENT: The eggs of the Little Crake are seven or eight in number. They are somewhat elliptical in shape, smooth in texture, yellowish-brown in ground colour, marbled and indistinctly blotched with olive-brown, and occasionally speckled with very dark brown. Occasionally the markings, instead of being uniformly distributed over the surface of the egg, are most numerous on the major half, or form an ill-defined cap on the larger end. Average measurement, 1·2 inch in length, by ·85 inch in breadth. Incubation is performed by the female, and lasts from twenty-one to twenty-four days.

DIAGNOSTIC CHARACTERS: The eggs of this Crake require careful identification, as they very closely resemble those of Baillon's Crake, only differing in being slightly larger.

Family COLUMBIDÆ.　　　　　　　　　　Genus TURTUR.

EASTERN TURTLE DOVE.

TURTUR ORIENTALIS (*Latham*).

(British: Very rare abnormal autumn migrant.)

Double Brooded. Laying season, May to August. (December to March, Central Provinces of India.)

BREEDING AREA: North-western Oriental region and South-eastern Palæarctic region. The breeding area of the Eastern Turtle Dove is most imperfectly defined, owing partly to the confusion existing between this species and several closely allied forms. The present species breeds in India from the Central Provinces northwards to the lower ranges of the Himalayas (4000 to 6000 feet), from Afghanistan to Sikhim; it is also known to do so in South-eastern Siberia, and possibly in Mongolia, Thibet, and Northern China. Whether it breeds in Japan is by no means clear.

BREEDING HABITS: In most parts of its northern area of dispersal the Eastern Turtle Dove is migratory, and even in the south is subject apparently to much local movement during the non-breeding season. Capt. Hutton states that it arrived in its summer quarters at Mussoorie in April, leaving again in October. In its habits it is not known to differ in any important respect from the nearly allied European Turtle Dove. It probably pairs for life, but makes a new nest for each brood. It cannot be regarded as gregarious during the breeding season, but numbers of pairs nest within comparatively small areas of suitable country, and as soon as the young are reared the birds begin to flock. Its favourite haunts are wooded mountain sides, pine forests, groves and clusters of trees. The note in the breeding season is described as a thrice-repeated guttural *coo*, and unlike that of other Doves. The nest is generally placed not far from the extremity of a horizontal branch, and is a circular, flat, mat-like structure of neatly-arranged twigs, the cavity in the centre being somewhat deep. Hume states that the nest is rather more substantial than that of many other Turtle Doves. Of the actions of the birds at the nest I find nothing of special interest recorded.

RANGE OF EGG COLOURATION AND MEASUREMENT: The eggs of the Eastern Turtle Dove are two in number, oval in form, glossy, and pure white. Average measurement, 1·2 inch in length, by ·92 inch in breadth. Incubation is performed by both sexes, and lasts about sixteen days.

DIAGNOSTIC CHARACTERS: I know of no character by which the eggs of the Eastern Turtle Dove can be distinguished from those of allied races and species. Locality is of some service in their identification, but great care is necessary, and the parents should be seen. It may be remarked that the eggs of the Common Turtle Dove are creamy-white, not pure white, as in the present species.

APPENDIX I.

LIST OF BRITISH SPECIES
WHOSE NESTS AND EGGS ARE AT PRESENT UNKNOWN TO SCIENCE.

Family FRINGILLIDÆ.   Genus EMBERIZA.
Sub-family EMBERIZINÆ.

RUSTIC BUNTING.

EMBERIZA RUSTICA, *Pallas.*

(British : Rare abnormal autumn migrant.)

The breeding grounds of this species are in sub-Arctic Europe and Asia, and extend from the Baltic in the west to Kamtschatka in the east. This Bunting does not appear to range further north in Europe than lat. 65°, and in Asia not beyond lat. 62°. The eggs of this species have never been thoroughly authenticated. Dresser describes eggs sent to him from the vicinity of Archangel; Professor Newton furnishes some particulars of another egg presumed to belong to this species; whilst Mr. Seebohm has described others from Archangel and the Altai; but in all these cases the identification is incomplete. Until the parents are obtained with the nest and eggs, we are fully justified in rejecting these specimens as unreliable.

Family LANIIDÆ.   Genus LANIUS.

PALLAS'S GRAY SHRIKE.

LANIUS MAJOR, *Pallas*.

(British : Fairly frequent autumn migrant.)

The breeding grounds of this species are in Siberia and Manchooria, south of about lat. 65°. The nest and eggs are still unknown to science.

Family TURDIDÆ.   Genus GEOCICHLA.
Sub-family *TURDINÆ*.

SIBERIAN GROUND THRUSH.

GEOCICHLA SIBIRICA (*Pallas*).

(British : Very rare abnormal autumn migrant.)

The breeding grounds of this species are presumed to be in the valleys of the Yenesay and the Lena, and in Japan. Nothing is known of the nest and eggs of this Thrush.

Family CYPSELIDÆ.   Genus CHÆTURA.

NEEDLE-TAILED SWIFT.

CHÆTURA CAUDACUTA (*Latham*).

(British : Very rare abnormal autumn migrant.)

The breeding grounds of this species are in South-eastern Siberia, Mongolia, the Eastern Himalayas, Thibet, North China, and Japan. Although some details of the nidification of this Swift in Mongolia are recorded by

Prjevalsky, the eggs are still unknown to science. It will be very interesting to learn whether these resemble in colour those of the typical Swifts. It is said to nest in cliffs and hollow trees, and to be social if not even gregarious.

Family CHARADRIIDÆ.   Genus TOTANUS.
Sub-family *TOTANINÆ*.

SOLITARY SANDPIPER.
TOTANUS SOLITARIUS (*Wilson*).

(British : Very rare abnormal autumn migrant.)

This species apparently breeds in the Northern United States, from about lat. 44° and up to the limits of forest growth near the Arctic Circle. Incredible as it may seem, the nest and eggs still remain unknown to science, for it is impossible to accept the description of the latter given by the late Dr. Brewer without authentication. There can be little doubt that this species lays its eggs in the deserted nests of other birds in low trees, like its Old World representative the Green Sandpiper is known to do. Search should be made in such places in the summer haunts of this species.

Family CHARADRIIDÆ.   Genus TRINGA.
Sub-family *SCOLOPACINÆ*.

SIBERIAN PECTORAL SANDPIPER.
TRINGA ACUMINATA (*Horsfield*).

(British : Very rare abnormal autumn migrant.)

The breeding grounds of this species are probably in Dauria, the Tchuski Land, and Kamtschatka. Nothing

NON-INDIGENOUS BRITISH BIRDS. 337

whatever is known of the habits of this species during the breeding season, or of its nest and eggs.

Family CHARADRIIDÆ. Genus TRINGA.
Sub-family SCOLOPACINÆ.

CURLEW SANDPIPER.

TRINGA SUBARQUATA (*Güldenstädt*).

(British : Fairly common spring and autumn coasting migrant ; few, winter.)

The breeding area of this species is almost unknown. Middendorff observed this species in summer on the Taimyr peninsula. Other nesting grounds may be probably on the Liakoff Islands, and on undiscovered lands in the North Polar Basin. To the present writer it seems by no means impossible that some individuals of this species may breed in the Antarctic regions. Nothing is yet known of its nest and eggs.

Family CHARADRIIDÆ. Genus TRINGA.
Sub-family SCOLOPACINÆ.

KNOT.

TRINGA CANUTUS, *Linnæus*.

(British : Common spring and autumn coasting migrant, especially the latter ; few, winter.)

The breeding grounds of this species are situated in the North Polar Basin, mostly, if not entirely, above lat. 80°. The Knot probably breeds on all suitable land within this area up to the Pole. That the bird is so rare in the north of continental Europe and Asia is strong presumptive evidence that no suitable land exists in the Polar Basin north of that area, and that its only breeding grounds are in Greenland and on the various islands that lie in high latitudes north of the American continent. Although many of the habits of this species during the breeding season have been observed, and its young in down secured, the eggs still remain undiscovered. It always seems to the present writer a most unpardonable and incredible piece of neglect on the part of the naturalist attached to the latest British Polar Expedition to have missed the eggs of the Knot. The bird was observed to arrive at its breeding places, to pair, and then actually to be lost sight of until the eggs were hatched! Several reputed eggs of the Knot are in collections, but none of them are authenticated. The reputed egg obtained by the Greely Expedition near Fort Conger is unidentified, and apparently too small (1·1 inch in length, by 1·0 inch in breadth). The egg in the possession of Mr. Seebohm (which I have examined), although unauthenticated, is more likely to be genuine so far as size is concerned, being similar to that of the Common Snipe, but paler in ground colour. This egg was obtained at Disco in Greenland ; in my opinion a locality much too far south. This, however, is not the most southerly locality at which reputed eggs of the Knot have been obtained. Mr. Raine, in his *Bird-nesting in North-west Canada*, figures and describes what he asserts to be two eggs of this bird, taken on the 20th of June, 1889, at Rædodavmsi, in Iceland ! The account is circumstantial enough, but unfortunately the parent birds appear not to have been obtained or even identified. It is only fair to say that Mr. Raine's eggs agree apparently in colour with that obtained by Lieutenant Greely, but are larger in size, and certainly, judging from the illustrations, very abnormal in appearance. The nest is described as a depression lined with bits of drift-weed; the eggs, as having the ground colour pale pea-green, finely speckled with ashy-brown : size 1·5 inch in length, by 1·0 inch in breadth (*op. cit.* p. 188, Pl. II., Figs. 1 and 2). Mr. Raine's collectors seem to have been fully aware of the importance of their discovery, and were too anxious to wait, after finding the nest with two eggs, for the full complement to be laid. I can only repeat that without authentication the eggs must be rejected by scientific naturalists as valueless. I might also remark that the Gray Phalarope breeds in Iceland, and that in nuptial plumage it bears a somewhat close resemblance to the Knot in breeding dress, both species having the under-parts rich chestnut during summer.

Family LARIDÆ.
Sub-family LARINÆ.

Genus RHODOSTETHIA.

ROSS'S GULL.

RHODOSTETHIA ROSEA, *Macgillivray*.

(British : Very rare nomadic winter migrant.)

The breeding grounds of this species are probably in the Polar regions lying north of lat. 75° or 80°. There is some evidence to suggest that one great breeding place of this Gull is situated either on the Liakoff Islands, or on undiscovered land lying to the north of Wrangel Island. Others probably occur on the Arctic Archipelago, north of Prince Albert Land. The egg reputed to be of this species taken in Greenland and forwarded to England from Disco is totally unauthenticated.

Family PROCELLARIIDÆ. Genus PUFFINUS.

GREAT SHEARWATER.

PUFFINUS MAJOR, *F. Faber.*

(British : Summer visitor during the period of the Antipodean winter.)

There can be no doubt whatever that this species is a bird of the Southern Hemisphere, and the fact, to my mind, is absolutely proved by the following circumstances. The nesting haunts of all Petrels that breed in the Northern seas are now fairly well known (especially in the Atlantic), but no resort of the Great Shearwater has been discovered. In the *Migration of Birds* I placed too much reliance upon Messrs. Baird, Brewer, and Ridgway's very circumstantial statement that this Shearwater bred in Greenland, but subsequent research has led me to reject it. This species evidently spends the period of the southern winter in the Northern Hemisphere, and after rearing its young in still unknown places in the Southern Seas, retires north to spend a second summer with us. The bird is well known in the North Atlantic during that period, and has been observed with great regularity to arrive at the fishing grounds off the coasts of New England and British North America in May, and to remain until October or November, when it retires to its home in the Southern Hemisphere to breed. Of the probable thousands of individuals of this species examined by Captain J. W. Collins, caught at these fishing grounds, *not one showed any traces of breeding!* Again, this Shearwater has been observed at Tierra del Fuego and off the Cape of Good Hope. The reason it has not been observed more widely and commonly in the Southern Seas is because it is collected in a few chosen resorts, and at this season is very nocturnal in its habits ; during its sojourn in the

North Atlantic it is more wandering in its habits, and spread over a wider area, as is the case with many other species. Lastly, and perhaps most significant fact of all, I have examined an example of this Shearwater from Greenland, still I believe in the collection of Mr. Hargitt, which is moulting its quills and other feathers on the 28th of June! This unquestionably confirms the suggestion that this species breeds in the Southern Hemisphere, and that it moults after the season of reproduction is over, in its winter quarters, as so many other birds are known to do. That the eggs of the Great Shearwater will eventually be found on some ocean islet or coast washed by the open Southern Seas amounts to an absolute certainty. To search for them north of the Equator is futile. The circumstance is quite in accord with our present knowledge of the Migration of Birds. Migration as a science is yet in its earliest infancy, and to that deplorable fact must be attributed the various erroneous statements that have been made concerning the geographical distribution of the Great Shearwater and many other species of Petrels. Unfortunately they are birds of nocturnal habits, especially at their breeding stations, and this to a great extent helps to keep our knowledge of their whereabouts so limited.

Family PROCELLARIIDÆ.   Genus ŒSTRELATA.

COLLARED PETREL.

ŒSTRELATA TORQUATA (*Macgillivray*).

(British : Very rare abnormal migrant.)

The late John Macgillivray, who discovered this Petrel on Aneiteum, one of the New Hebrides, and who states that it is also found on Tanna and Erromango, in the same group of islands, found it breeding in burrows on the wooded tops of mountains in the interior of the islands, but unfortunately failed to obtain eggs. A young chick covered with black down was brought to him on the 14th of February. This Petrel perhaps breeds too near to the Equator to have any very regular or extended normal migration north, and it can only be looked upon as a very rare straggler to the Northern Seas. It is cause for surprise that the eggs are still unknown to science. There can be little doubt that this Petrel also breeds on the Fiji Islands, as it was obtained on Fiji in 1878 by Kleinschmit. The laying season is probably December and January.

Family PROCELLARIIDÆ. Genus ŒSTRELATA.

CAPPED PETREL.

ŒSTRELATA HÆSITATA (*Kuhl*).

(British : Very rare abnormal migrant.)

According to Mr. Salvin, our highest authority on this family of birds, the home of the Capped Petrel is on the Windward Islands, some of the most southerly of the West Indies. It probably also breeds on various islets off the coast of Venezuela, but at present nothing whatever is known of its nest and eggs, whilst its exact breeding area still remains undefined. Breeding as it most certainly does so near to the Equator, its migrations are necessarily very restricted, and its appearance in our seas purely abnormal.

Family PROCELLARIID.E. Genus DAPTION.

CAPE PETREL.

DAPTION CAPENSE (*Linnæus*).

(British : Very rare straggler, but doubtless a common visitor to the Northern Seas during the period of winter in the Southern Hemisphere.)

This Petrel is another Southern Hemisphere species observed in abundance in most of the Southern Seas. It is said to breed on the island of South Georgia, and doubtless does so on many other ocean islands in this region, but its eggs still remain undescribed. When we know so little of the area inhabited by this species in summer, it may be rash to state that its appearance in the Northern Seas is thoroughly normal; but in the face of what we do know respecting the laws which govern the migrations of the Petrels, it seems a little premature to say that there is no "adequate reason for including this species among the birds of Great Britain" [the only specimen observed in our area is an *Irish* one, so that the remark is all the more unhappy] "or even of Europe, for its home is essentially the Southern Hemisphere" (*Manual of British Birds*, p. 714). We might just as well reject the Sooty Shearwater, the Great Shearwater, and Wilson's Petrel for the same reason. It may be that the northern flights of the Cape Petrel extend to the Indian and North Pacific Oceans rather than to the North Atlantic Ocean, but this is a matter of detail and quite beside the argument.

Family COLYMBIDÆ.   Genus COLYMBUS.

WHITE-BILLED DIVER.

COLYMBUS ADAMSI, *Gray.*

(British : rare nomadic winter migrant.)

The breeding grounds of this Diver are probably circumpolar, and confined to the Arctic regions of both hemispheres. The northern limit is not yet determined. MacFarlane states that this species abounds during the breeding season in Franklin and Liverpool Bays, on the coasts of Arctic America, and was also met with occasionally on the lakes in the interior, but he failed to obtain any authenticated eggs. The two eggs referred to by Messrs. Baird, Brewer, and Ridgway in their *Water Birds of North America*, ii. p. 452, as those of this species, may have belonged to the Great Northern Diver, as suggested by MacFarlane. No thoroughly identified eggs are yet known to science.

APPENDIX II.

LIST OF SPECIES WHOSE CLAIM TO RANK AS BRITISH
IS DOUBTFUL.

IT is a matter of very great difficulty in many cases to decide whether a species has sufficient claim to rank as "British" or not. As my readers know, I am more apt to err on the side of inclusion than on that of omission, and I am led to do this partly through my long study of the migration flight of birds. In a great many cases I admit that it is simply wonderful how individuals of a species have managed to wander so far from their normal areas of dispersal as the British Archipelago; yet when we study their usual migrations in conjunction with their habitat, much of the wonder is apt to vanish. In my opinion some species have been most unfairly expunged from the British list; others have been included with perhaps too little cause. Sedentary species should only be admitted on the clearest possible evidence; birds that breed in the southern Tropics or south temperate zones are even less likely to wander to our islands, because their migrations are in every known instance remarkably restricted; aquatic species are more likely to travel long distances than terrestrial species. Species that are kept regularly and extensively in confinement should always be regarded with grave suspicion. On the other hand, many birds of migratory habits breeding to the east or north of us are very likely to wander to us occasionally; whilst birds whose range does not extend so far north as our area are in some cases apt to overshoot their mark and reach it, but the date must in all probability be in spring. That some American species have managed to reach our islands we are bound to believe, especially water birds; for we know that vast numbers of Knots do the double passage every year, *viâ* Greenland, Iceland, and the Faroes. On the other hand, it is possible that many stray northern Nearctic birds reach Western Europe by way of Asia. The following is a list of such species, individuals of which have been alleged to occur within the limits of the British Islands. With the exception of those species marked by an asterisk, I do not think any possible claims have been ignored, and that they are wisely excluded for reasons given under each species. Those so marked I consider the evidence is not quite sufficiently strong for their inclusion as "British," but the probability is that future evidence of a stronger and more unimpeachable character may be obtained which will ultimately win for some of them the honour of a place. Scarcely a year passes without some bird new to the British avi-fauna reaching our islands. It therefore behoves the fortunate possessors of these rare stragglers to have their specimens properly examined by competent authorities, to see that all the data concerning them are collected, and by this means to preclude the possibility of future doubt being cast over them.

ALPINE CHOUGH.

PYRRHOCORAX ALPINUS, *Koch*.

Has once occurred. A sedentary species, and known to have been kept in confinement in our islands. The individual probably escaped, although the date of its capture (April) is one point in its favour. *Habitat:* Mountains of Central and Southern Europe and Asia, as far east as North-western China. It is recorded from Heligoland.

* RED-WINGED STARLING.

AGELÆUS PHŒNICEUS (*Linnæus*).

A dozen or more British records. A migratory species, but unfortunately one often kept in confinement and imported freely. It is possible that some of the individuals may have been abnormal migrants, but the above fact taints their record. *Habitat:* North America, up to about lat. 62°.

* RUSTY GRAKLE.

SCOLECOPHAGUS FERRUGINEUS, *Gmelin*.

Has once occurred. A migratory species, but as it is frequently kept in confinement it is possible that the individual had escaped from captivity. I have examined this specimen, and must in common fairness say that it bears no trace of having been in a cage. *Habitat:* Arctic regions of America, north to the limits of forest growth.

* MEADOW STARLING.

STURNELLA MAGNA, *Linnæus*.

Three British records. A migratory species, but one often kept in confinement, so that these individuals may probably be escapes. *Habitat:* Eastern United States.

GOLD-VENTED BULBUL.

PYCNONOTUS CAPENSIS (*Linnæus*).

Has once occurred. Practically a sedentary species. Probably an escape, the date of its capture (January) and the area it inhabits being utterly opposed to any migratory movement. *Habitat:* South Africa; apparently confined to Cape Colony.

SOUTH AFRICAN SERIN.

SERINUS CANICOLLIS (*Swainson*).

Two British records. Practically a sedentary species. Certainly escaped individuals. *Habitat:* South Africa.

YELLOW-RUMPED SEED-EATER.

SERINUS ICTERUS (*Bonn et Vieillot*).

One British record. Practically a sedentary species. Certainly an escape. *Habitat:* West Africa.

NONPAREIL FINCH.

CYANOSPIZA CIRIS (*Linnæus*).

One British record. A species subject to some migratory movement, but, as was suggested long ago by Montagu, probably an escape. *Habitat:* Central and North America.

* WHITE-THROATED SPARROW.

ZONOTRICHIA ALBICOLLIS (*Gmelin*).

Two British records. A migratory species, and the date of capture (autumn) is a point in favour of the individuals having reached our area by abnormal migration. *Habitat:* North America.

* BRANDT'S SIBERIAN BUNTING.

EMBERIZA CIOIDES CASTANEICEPS, *Moore*.

One British record. A partially migratory species, and the date (October) and locality (near Flamborough) of the specimen are strong points in favour of the supposition that it was an abnormal migrant from the far east. The specimen is said to approach most closely to the Chinese, and not to the Siberian race of this species. *Habitat:* Northern China.

* RUBY-CROWNED WREN.

REGULUS CALENDULA, *Linnæus.*

One British record. A species subject to some migratory movement. The date of capture (summer) is a point against the individual occurring in a wild state; but on the other hand the species has been known to stray to Greenland, which fact is certainly in favour of an individual prolonging its abnormal flight to our shores. The fact that it remained unidentified for six years—it is a very strongly marked species—also tells adversely against its bonâ-fides. *Habitat:* North America.

* AMERICAN ROBIN.

TURDUS MIGRATORIUS, *Linnæus.*

Apparently two British records. A migratory species, and probably the examples obtained were on abnormal flight, seeing that the bird has also been captured on Heligoland, in a very exhausted condition. Unfortunately it is a species often kept in confinement. These stray individuals may have reached us *viâ* Asia. *Habitat:* North America up to the Arctic regions.

WHITE-COLLARED FLYCATCHER.

MUSCICAPA COLLARIS, *Bechstein.*

Apparently one British record. The late Mr. Gould included this species as British, but on insufficient evidence. *Habitat:* Southern Europe.

* RED-RUMPED SWALLOW.

HIRUNDO RUFULA, *Temminck*.

One British record. Mr. Rodd states that an example of this species was seen by him at Penzance. This evidence is not sufficient to allow of the species claiming a place in the British list. It is a migratory bird. *Habitat:* South-eastern Europe, Asia Minor, and Palestine, in summer; Eastern Africa in winter. It has been recorded from Heligoland.

* AMERICAN TREE SWALLOW.

TACHYCINETA BICOLOR (*Vieillot*).

One British record. A migratory species, and visiting the Arctic regions of America in summer. I see nothing improbable in a stray individual reaching our shores *via* Greenland, Iceland, and the Faroes, or even by way of Asia. *Habitat:* North America up to the Arctic regions in summer; West Indies, Central and South America in winter.

BLUE-TAILED BEE-EATER.

MEROPS PHILIPPINUS, *Linnæus*.

One British record. A sedentary species inhabiting an area from which it is impossible to believe any individual could by the remotest chance reach our islands unaided by man. Probably an escape, or a foreign skin passed off as British. *Habitat:* South-eastern Asia from India and South China to the Malay Archipelago.

ABYSSINIAN ROLLER.

CORACIAS LEUCOCEPHALUS, *P.L.S. Müller.*

Two British records. A sedentary species. The history of the two examples reputed to have been captured in our islands is utterly unreliable, and we may dismiss the species without further comment. *Habitat:* Arabia and Africa south of the Desert.

* INDIAN ROLLER.

CORACIAS INDICUS, *Linnæus.*

One recent British record. A partially migratory species, and according to the date of capture (October), the individual in question very probably reached our islands on abnormal flight. *Habitat:* Persia, Afghanistan, and India.

JUGGER FALCON.

FALCO JUGGER, *Gray.*

One British record. A sedentary species, the individual obtained in our islands most probably escaping from captivity. *Habitat:* India.

* AMERICAN KESTREL.

FALCO SPARVERIUS, *Linnæus.*

One British record. A migratory species, but unfortunately the date of capture (May) is against the possibility of abnormal flight, unless the individual in question had resided with us from the previous autumn. The evidence of capture is far from satisfactory. *Habitat:* North America up to the Arctic regions.

BLACK-WINGED KITE.

ELANUS CŒRULEUS, *Desfontaines.*

One British record. A species said to be subject to some migratory movement. *Habitat:* Tropical and sub-tropical Africa. The example in question may have been an imported skin, but the species has occurred accidentally elsewhere in Europe.

* TRUMPETER SWAN.

CYGNUS BUCCINATOR, *Richardson.*

One British record (four examples). A migratory species. The date of capture (October) is a point in favour. The evidence of identification is, however, unreliable, and in the instance of one individual (probably in all by inference) points to absolute error. *Habitat:* North America, from lat. 42° northwards to the Arctic regions. Breeds freely in confinement, and has long been naturalized in our islands.

* AMERICAN SWAN.

CYGNUS AMERICANUS, *Sharpless.*

Two British records (one of five individuals). A migratory species, and judging from the dates of capture

(February and December) and the locality (Scotland), the individuals in question probably reached our area by abnormal flight. The identification is somewhat doubtful. *Habitat:* Arctic and sub-Arctic America.

* BAR-HEADED GOOSE.
ANSER INDICUS (*Latham*).

Several British records. A migratory species; but as it is one that is frequently kept in semi-captivity on ornamental waters, it is difficult to say whether the individuals recorded were escaped birds or not. *Habitat:* Mongolia, wintering in India.

* CHINESE GOOSE.
ANSER CYGNOIDES (*Linnæus*).

Several British records. A migratory species, but kept largely on ornamental waters, a fact which at once casts doubt on the individuals obtained. It is far from improbable that stray individuals may reach us on abnormal flight, carried with the great east to west wave of Palæarctic migration. *Habitat:* East Siberia in summer; China in winter.

* CANADA GOOSE.
BERNICLA CANADENSIS (*Linnæus*).

Several British records. A migratory species, but the individuals obtained are always open to the suggestion that they are escaped birds, the species being largely kept on ornamental waters. Personally I am inclined to think that stray examples may reach us on abnormal flight. *Habitat:* Arctic and sub-Arctic America.

EGYPTIAN GOOSE.

CHENALOPEX ÆGYPTIACA (*Linnæus*).

Several British records. A sedentary species. There can be no doubt whatever that these records in every case refer to escaped birds, the species being largely kept in captivity. *Habitat:* East and South Africa.

SPUR-WINGED GOOSE.

PLECTROPTERUS GAMBENSIS (*Linnæus*).

Several British records. A sedentary species. Precisely the same remarks apply as to the preceding species. It is widely kept in captivity, and its habits and geographical area are both opposed to any abnormal flight to our area. *Habitat:* West and South Africa.

* RING-NECKED DUCK.

FULIGULA COLLARIS (*Donovan*).

One British record. It seems hard to refuse a place to a species which was first described by Donovan in his work on British Birds from an individual obtained in our islands, but as the specimen in question was purchased in Leadenhall Market, and as no other examples have been secured, the matter is perhaps best left in abeyance. There can be little if any doubt that this individual came from its usual haunts on abnormal flight, but there is just the possibility that it was imported with other wild-fowl from the Continent. We must, however, bear in mind that in Donovan's day (1801) the means of transit were small. The bird has far more right to a place in the British list than many others whose position, sanctioned solely by custom, seems unassailable. *Habitat:* Canada and the Northern States in summer; Southern States, Central America, and West Indies in winter.

LITTLE GREEN HERON.

BUTORIDES VIRESCENS (*Linnæus*).

One British record. A migratory species, but one whose range scarcely extends far enough north to render an abnormal visit to our islands possible. I find that Mr. Seebohm admits this species into his recently-published list of British birds as an "accidental visitor," but does not even mention the preceding species. It is difficult to explain such capricious treatment, and I leave the reader to say which of the two has the most right even to a bare allusion. *Habitat:* North America; migratory in South Canada and the Northern States; resident further south.

SOUDAN CRANE.
GRUS PAVONINA (*Linnæus*).

One British record. Said to be a sedentary species. There can be no doubt that the individual recorded was an escape, the date of capture (September) and the area inhabited precluding any possibility of abnormal flight. *Habitat:* Central and West Africa.

ANDALUCIAN HEMIPODE.
TURNIX SYLVATICA (*Desfontaines*).

Three British records. A sedentary species. There can be no doubt that the individuals in question had been imported, and were either turned out or had escaped. *Habitat:* South-western Europe, and North-western Africa.

* MARSH SANDPIPER.
TOTANUS STAGNATILIS, *Bechstein*.

One British record. A migratory species. I am informed by the Hon. Walter Rothschild that the reputed example obtained at Tring reservoir has been lost or mislaid. I see no reason why stray individuals of this Sandpiper should not reach our area, but for the present it is perhaps the wisest course to exclude the species from the British list. *Habitat:* Southern Europe, North Persia, Southern Turkestan, and Southern Siberia in summer; Africa, India, Burma, the Malay Archipelago, and Australia in winter. An example has been obtained on Heligoland.

BRIDLED TERN.
STERNA ANÆSTHETA, *Scopoli*.

One British record. It is difficult to believe that the individual in question reached our area on abnormal flight. The date of capture (September) is dead against it; the area it inhabits is also one from which it is highly improbable that it would wander to our shores. The species is very rightly excluded from the British list. *Habitat*: Oceanic tropical zone. I have only included the Sooty Tern and the Noddy Tern as British species out of deference to their long standing on the list. They may have occurred here on abnormal flight, but the circumstance is very exceptional, and I should like to see them expunged.

PIED-BILLED GREBE.
PODILYMBUS PODICEPS (*Linnæus*).

One British record. A stray example of this species might reach our islands, seeing that individuals have occurred on the Bermudas, but there are circumstances in the case which seem to suggest error, and I think the bird is very rightly ignored. *Habitat:* America.

* CAROLINA RAIL.
CREX CAROLINA (*Linnæus*).

One British record. A migratory species. What I said of this example in *Game Birds and Wild Fowl* may very aptly be repeated here. " Naturalists for some inscrutable reason decline to admit the Carolina Crake as an established British species, but the known wandering habits of birds of this family, in addition to the fact of its occurrence in Greenland, is strong evidence in its favour of having reached our islands voluntarily." *Habitat:* Northern United States and Canada, as far north as lat. 62°, in summer; Southern States, Mexico, Central America, and the West Indies in winter.

PURPLE GALLINULE.
PORPHYRIO CŒRULEUS (*Vandelli*).

Several British records. A sedentary species. There can be no doubt that the individuals recorded had escaped from confinement. *Habitat:* Italy, Spain, and North-west Africa.

GREEN-BACKED GALLINULE.
PORPHYRIO SMARAGDONOTUS, *Temminck*.

Several British records. A sedentary species. There can be no doubt that all the individuals obtained in our islands are escaped birds. *Habitat:* Africa, with the exception of the North-west, where it is replaced by the preceding species.

MARTINIQUE GALLINULE.

PORPHYRIO MARTINICUS (*Linnæus*).

Two British records. A sedentary species. There can be no doubt that the individuals in question had escaped from captivity, perhaps from passing ships. *Habitat:* Tropical and sub-tropical America.

* PASSENGER PIGEON.

ECTOPISTES MIGRATORIUS (*Linnæus*).

Several British records. A migratory species, and one which could probably cross the Atlantic, but unfortunately the species is largely kept in confinement, and has even been turned loose in our islands, so that the individuals recorded are subject to the gravest doubt. *Habitat:* North America, North-eastern States, and Canada, up to lat. 65°, in summer; United States in winter.

INDEX.

Abyssinian Roller, 352
Accentor, Alpine, 88
Accentor alpinus, 88
Acrocephalus aquaticus, 67
Acrocephalus turdoides, 65
Aëdon galactodes, 63
Ægialitis hiaticula, 227
Ægialitis minor, 229
Ægialitis vocifera, 226
Ægialophilus asiaticus, 231
Agelæus phœniceus, 347
Alca impennis, 314
Alpine Accentor, 88
Alpine Chough, 347
Alpine Pipit, 39
American Bittern, 200
American Golden Plover, 236
American Goshawk, 142
American Kestrel, 352
American Pectoral Sandpiper, 266
American Robin, 350
American Stint, 270
American Swan, 353
American Teal, 163
American Tree Swallow, 351
American White-winged Crossbill, 6
American Widgeon, 161
Ampelis garrulus, 51
Anas americana, 161
Anas carolinensis, 163
Anas discors, 164
Andalucian Hemipode, 357
Anous stolidus, 298
Anser albifrons, 151
Anser albifrons minutus, 153
Anser brachyrhynchus, 150
Anser cygnoides, 354
Anser indicus, 354

Anser segetum, 148
Anthus campestris, 41
Anthus cervinus, 45
Anthus richardi, 43
Anthus spipoletta, 39
Aquatic Warbler, 67
Aquila nævia, 133
Archibuteo lagopus, 139
Arctic blue-throated Robin, 79
Ardea alba, 209
Ardea bubulcus, 204
Ardea comata, 206
Ardea garzetta, 207
Ardea purpurea, 210
Ardetta minuta, 199
Asiatic Golden Plover, 234
Astur atricapillus, 142
Astur palumbarius, 141
Athene noctua, 110
Auk, Great, 314
Auk, Little, 313
Avocet, Common, 239

Bar-headed Goose, 354
Bar-tailed Godwit, 256
Barred Warbler, 61
Bartram's Sandpiper, 245
Bean Goose, 148
Bee-eater, 103
Bee-eater, Blue-tailed, 351
Belted Kingfisher, 107
Bernacle Goose, 157
Bernicla brenta, 154
Bernicla canadensis, 354
Bernicla glaucogaster, 156
Bernicla leucopsis, 157
Bernicla ruficollis, 158
Bewick's Swan, 145
Bittern, American, 200
Bittern, Little, 199

Black-bellied Dipper, 89
Black-billed Cuckoo, 96
Black-headed Bunting, 23
Black-headed Gull, Great, 290
Black-headed Gull, Mediterranean, 291
Black Kite, 135
Black-necked Grebe, 324
Black Stork, 197
Black-tailed Godwit, 254
Black Tern, 308
Black-throated Ouzel, 78
Black-throated Wheatear, 83
Black-winged Kite, 353
Blue-tailed Bea-eater, 351
Blue-throated Robin, Arctic, 79
Blue-winged Teal, 164
Bonaparte's Gull, 293
Bonaparte's Sandpiper, 261
Botaurus lentiginosus, 200
Brambling, 16
Brandt's Siberian Bunting, 349
Brent Goose, 154
Brent Goose, White-bellied, 156
Bridled Tern, 358
Broad-billed Sandpiper, 264
Brunnich's Guillemot, 311
Bubo maximus, 117
Buff-backed Heron, 204
Buff-breasted Sandpiper, 276
Buffel-headed Duck, 178
Buffon's Skua, 282
Bulbul, Gold-vented, 348
Bulwer's Petrel, 316
Bulweria columbina, 316
Bunting, Black-headed, 23
Bunting, Lapland, 21
Bunting, Little, 27
Bunting, Ortolan, 25
Bunting, Rustic, 334
Bustard, Great, 216
Bustard, Little, 218
Bustard, Macqueen's, 219
Butarides virescens, 356
Buzzard, Rough-legged, 139

Calandra Lark, 31
Calandrella brachydactyla, 35
Calcarius lapponicus, 21
Canada Goose, 354

Canary, 13
Cape Petrel, 343
Capped Petrel, 342
Caprimulgus ægyptius, 102
Caprimulgus ruficollis, 101
Carolina Rail, 359
Carpodacus erythrinus, 11
Caspian Sand Plover, 231
Caspian Tern, 302
Ceryle alcyon, 107
Chætura caudacuta, 335
Charadrius americanus, 236
Charadrius fulvus, 234
Charadrius helveticus, 232
Chen hyperboreus, 147
Chenalopex ægyptiaca, 355
Chinese Goose, 354
Chough, Alpine, 347
Ciconia alba, 195
Ciconia nigra, 197
Cinclus aquaticus melanogaster, 89
Clangula albeola, 178
Clangula glaucion, 179
Coccystes glandarius, 93
Coccyzus americanus, 97
Coccyzus erythrophthalmus, 96
Collared Petrel, 341
Colymbus adamsi, 344
Colymbus glacialis, 322
Common Avocet, 239
Common Crane, 212
Common Pratincole, 223
Common Stilt, 237
Coracias garrulus, 105
Coracias indicus, 352
Coracias leucocephalus, 352
Courser, Cream-coloured, 221
Crake, Little, 330
Crane, Common, 212
Crane, Demoiselle, 214
Crane, Soudan, 357
Cream-coloured Courser, 221
Creeper, Wall-, 47
Crested Lark, 37
Crex carolina, 359
Crex parva, 330
Crossbill, American White-winged, 6
Crossbill, European White-winged, 7

INDEX. 363

Cuckoo, Black-billed, 96
Cuckoo, Great Spotted, 93
Cuckoo, Yellow-billed, 97
Curlew Sandpiper, 337
Cursorius gallicus, 221
Cyanospiza ciris, 349
Cygnus americanus, 353
Cygnus bewicki, 145
Cygnus buccinator, 353
Cygnus musicus, 144
Cypselus melba, 99

Daption capense, 343
Demoiselle Crane, 214
Desert Wheatear, 84
Dipper, Black-bellied, 89
Diver, Great Northern, 322
Diver, White-billed, 344
Dove, Eastern Turtle, 331
Duck, Buffel-headed, 178
Duck, Harlequin, 171
Duck, Long-tailed, 173
Duck, Ring-necked, 355
Dusky Redshank, 252
Dusky Shearwater, 320

Eagle Owl, 117
Eagle, Spotted, 133
Eastern Turtle Dove, 331
Ectopistes migratorius, 360
Egret, Great White, 209
Egret, Little, 207
Egyptian Goose, 355
Egyptian Nightjar, 102
Egyptian Vulture, 121
Eider, King, 183
Eider, Steller's, 181
Elanoides furcatus, 137
Elanus cœruleus, 353
Emberiza cioides castaniceps, 349
Emberiza hortulana, 25
Emberiza melanocephala, 23
Emberiza pusilla, 27
Emberiza rustica, 334
Ereunetes griseus, 257
Erithacus suecica, 79
Eskimo Wimbrel, 241
European White-winged Crossbill, 7

Falco cenchris, 131

Falco jugger, 352
Falco sparverius, 352
Falco vespertinus, 129
Falcon, Jugger, 352
Fieldfare, 73
Finch, Nonpariel, 349
Finch, Scarlet Rose, 11
Finch, Serin, 14
Firecrest, 49
Flamingo, 188
Flycatcher, Red-breasted, 90
Flycatcher, White-collared, 350
Fringilla montifringilla, 16
Fuligula collaris, 355
Fuligula fusca, 175
Fuligula glacialis, 173
Fuligula histrionica, 171
Fuligula marila, 169
Fuligula nyroca, 167
Fuligula perspicillata, 176
Fuligula rufina, 166

Galerita Cristata, 37
Gallinule, Green-backed, 359
Gallinule, Martinique, 360
Gallinule, Purple, 359
Geocichla sibirica, 335
Geocichla varia, 71
Glareola pratincola, 223
Glaucous Gull, 191
Glossy Ibis, 191
Godwit, Bar-tailed, 256
Godwit, Black-tailed, 254
Gold-vented Bulbul, 348
Golden-eye, 179
Golden Plover, American, 236
Golden Plover, Asiatic, 234
Goose, Bar-headed, 354
Goose, Bean, 148
Goose, Bernacle, 157
Goose, Brent, 154
Goose, Brent, White-bellied, 156
Goose, Canada, 354
Goose, Chinese, 354
Goose, Egyptian, 355
Goose, Lesser Snow, 147
Goose, Lesser White-fronted, 153
Goose, Pink-footed, 150
Goose, Red-breasted, 158
Goose, Snow, Lesser, 147

Goose, Spur-winged, 355
Goose, White-bellied Brent, 156
Goose, White-fronted, 151
Goose, White-fronted, Lesser, 153
Goshawk, 141
Goshawk, American, 142
Grakle, Rusty, 347
Gray Phalarope, 243
Gray Plover, 232
Gray Shrike, Great, 55
Gray Shrike, Lesser, 53
Gray Shrike, Pallas's, 335
Great Auk, 314
Great Black-headed Gull, 290
Great Bustard, 216
Great Gray Shrike, 55
Great Northern Diver, 322
Great Reed Warbler, 65
Great Shearwater, 340
Great Snipe, 278
Great Spotted Cuckoo, 93
Great White Egret, 209
Grebe, Black-necked, 324
Grebe, Pied-billed, 358
Grebe, Red-necked, 328
Grebe, Sclavonian, 326
Green-backed Gallinule, 359
Green Heron, Little, 356
Greenland Redpole, 20
Green Sandpiper, 248
Griffon Vulture, 119
Grosbeak, Pine-, 9
Ground Thrush, Siberian, 335
Ground Thrush, White's, 71
Grus communis, 212
Grus pavonia, 357
Grus virgo, 214
Guillemot, Brunnich's, 311
Gull, Bonaparte's, 293
Gull, Glaucus, 288
Gull, Great Black-headed, 290
Gull, Iceland, 287
Gull, Ivory, 285
Gull, Little, 295
Gull, Mediterranean Black-headed, 291
Gull, Ross's, 339
Gull, Sabine's, 297
Gull-billed Tern, 304
Gyps fulvus, 119

Harlequin Duck, 171
Hawk Owl, 114
Hemipode, Andalucian, 357
Heron, Buff-backed, 204
Heron, Little Green, 356
Heron, Night, 202
Heron, Purple, 210
Heron, Squacco, 206
Hierofalco candicans, 123
Hierofalco gyrfalco, 127
Hierofalco islandus, 125
Himantopus melanopterus, 237
Hirundo rufula, 351
Hobby, Orange-legged, 129
Hooded Merganser, 185
Hooper Swan, 144
Hydrochelidon hybrida, 306
Hydrochelidon leucoptera, 310
Hydrochelidon nigra, 308
Hypolais hypolais, 69

Ibis, Glossy, 191
Iceland Gull, 287
Iceland Jer-Falcon, 125
Icterine Warbler, 69
Indian Roller, 352
Isabelline Wheatear, 81
Ivory Gull, 285

Jack Snipe, 280
Jer-Falcon, Iceland, 125
Jer-Falcon, Scandinavian, 127
Jer-Falcon, White, 123
Jugger Falcon, 352

Kestrel, American, 352
Kestrel, Lesser, 131
Killdeer Plover, 226
King Eider, 183
Kingfisher, Belted, 107
Kite, Black, 135
Kite, Black-winged, 353
Kite, Swallow-tailed, 137
Knot, 337

Lanius excubitor, 55
Lanius major, 335
Lanius minor, 53
Lapland Bunting, 21
Lapwing, Sociable, 225

Lark, Calandra, 31
Lark, Crested, 37
Lark, Shore-, 29
Lark, Short-toed, 35
Lark, White-winged, 33
Larus glaucus, 288
Larus ichthyaëtus, 290
Larus leucopterus, 287
Larus melanocephalus, 291
Larus minutus, 295
Larus philadelphia, 293
Lesser Gray Shrike, 53
Lesser Kestrel, 131
Lesser Snow Goose, 147
Lesser White-fronted Goose, 153
Limosa melanura, 254
Limosa rufa, 256
Linota linaria, 18
Linota linaria hornemanni, 20
Little Auk, 313
Little Bittern, 199
Little Bunting, 27
Little Bustard, 218
Little Crake, 330
Little Egret, 207
Little Green Heron, 356
Little Gull, 295
Little Owl, 110
Little Ringed Plover, 229
Little Stint, 267
Long-tailed Duck, 173
Loxia bifasciata, 7
Loxia enucleator, 9
Loxia leucoptera, 6

Macqueen's Bustard, 219
Marsh Sandpiper, 357
Martin, Purple, 92
Martinique Gallinule, 360
Meadow Starling, 348
Mealy Redpole, 18
Mediterranean Black-headed Gull, 291
Melanocorypha calandra, 31
Melanocorypha sibirica, 33
Merganser, Hooded, 185
Mergulus alle, 313
Mergus albellus, 186
Mergus cucullatus, 185
Merops apiaster, 103

Merops philippinus, 351
Merula atrigularis, 78
Milvus ater, 135
Monticola saxatilis, 86
Muscicapa collaris, 350
Muscicapa parva, 90

Needle-tailed Swift, 335
Neophron percnopterus, 121
Night Heron, 202
Nightjar, Egyptian, 102
Nightjar, Red-necked, 101
Noddy Tern, 298
Nonpareil Finch, 349
Northern Diver, Great, 322
Nucifraga caryocatactes, 1
Numenius borealis, 241
Nutcracker, 1
Nyctala tengmalmi, 109
Nyctea nyctea, 112
Nycticorax griseus, 202

Oceanites wilsoni, 317
Œstrelata hæsitata, 342
Œstrelata torquata, 341
Orange-legged Hobby, 129
Orphean Warbler, 59
Ortolan Bunting, 25
Otis macqueeni, 219
Otis tarda, 216
Otis tetrax, 218
Otocoris alpestris, 29
Ouzel, Black-throated, 78
Owl, Eagle, 117
Owl, Hawk, 114
Owl, Little, 110
Owl, Scops, 116
Owl, Snowy, 112
Owl, Tengmalm's, 109

Pagophila eburnea, 285
Pallas's Gray Shrike, 335
Passenger Pigeon, 360
Pastor roseus, 4
Pectoral Sandpiper, American, 266
Pectoral Sandpiper, Siberian, 336
Petrel, Bulwer's, 316
Petrel, Cape, 343
Petrel, Capped, 342

Petrel, Collared, 316
Petrel, Wilson's, 317
Phalarope, Gray, 243
Phalaropus fulicarius, 243
Phœnicopterus roseus, 188
Phylloscopus superciliosus, 57
Pied-billed Grebe, 358
Pigeon, Passenger, 360
Pine-Grosbeak, 9
Pink-footed Goose, 150
Pipit, Alpine, 39
Pipit, Red-throated, 45
Pipit, Richard's, 43
Pipit, Tawny, 41
Platalea leucorodia, 193
Plectopterus gambensis, 355
Plegadis falcinellus, 191
Plover, American Golden, 236
Plover, Asiatic Golden, 234
Plover, Caspian Sand, 231
Plover, Gray, 232
Plover, Killdeer, 226
Plover, Little Ringed, 229
Plover, Ringed, 227
Pochard, Red-crested, 166
Pochard, White-eyed, 167
Podiceps cornutus, 326
Podiceps nigricollis, 324
Podiceps rubricollis, 328
Podilymbus podiceps, 358
Pomatorhine Skua, 283
Porphyrio cœruleus, 359
Porphyrio martinicus, 360
Porphyrio smaragdonotus, 359
Pratincole, Common, 223
Progne purpurea, 92
Puffinus griseus, 319
Puffinus major, 340
Puffinus obscurus, 320
Purple Gallinule, 359
Purple Heron, 210
Purple Martin, 92
Purple Sandpiper, 262
Pycnonotus capensis, 348
Pyrrhocorax alpinus, 347

Rail, Carolina, 359
Recurvirostra avocetta, 239
Red-breasted Flycatcher, 90
Red-breasted Goose, 158

Red-breasted Snipe, 257
Red-crested Pochard, 166
Red-necked Grebe, 328
Red-necked Nightjar, 101
Redpole, Greenland, 20
Redpole, Mealy, 18
Red-rumped Swallow, 351
Redshank, Dusky, 252
Red-throated Pipit, 45
Redwing, 75
Red-winged Starling, 347
Reed Warbler, Great, 65
Regulus calendula, 350
Regulus ignicapillus, 49
Rhodostethia rosea, 339
Richard's Pipit, 43
Ring-necked Duck, 355
Ringed Plover, 227
Ringed Plover, Little, 229
Robin, American, 350
Robin, Arctic Blue-throated, 79
Rock Thrush, 86
Roller, 105
Roller, Abyssinian, 352
Roller, Indian, 352
Rose-coloured Starling, 4
Rose Finch, Scarlet, 11
Ross's Gull, 339
Rough-legged Buzzard, 139
Ruby-crowned Wren, 350
Ruddy Sheldrake, 160
Rufous Warbler, 63
Rustic Bunting, 334
Rusty Grakle, 347

Sabine's Gull, 297
Sanderling, 274
Sandpiper, American Pectoral, 266
Sandpiper, Bartram's, 245
Sandpiper, Bonaparte's, 261
Sandpiper, Broad-billed, 264
Sandpiper, Buff-breasted, 276
Sandpiper, Curlew, 337
Sandpiper, Green, 248
Sandpiper, Marsh, 357
Sandpiper, Purple, 262
Sandpiper, Siberian Pectoral, 336
Sandpiper, Solitary, 336
Sandpiper, Spotted, 246

Sandpiper, Yellow-legged, 250
Sand Plover, Caspian, 231
Saxicola deserti, 84
Saxicola isabellina, 81
Saxicola stapazina, 83
Scandinavian Jer-Falcon, 127
Scarlet Rose Finch, 11
Scaup, 169
Sclavonian Grebe, 326
Scolecaphagus ferrugineus, 347
Scolopax gallinula, 280
Scolopax major, 278
Scops Owl, 116
Scops scops, 116
Scoter, Surf, 176
Scoter, Velvet, 175
Seed-eater, Yellow-rumped, 348
Serin Finch, 14
Serin, South African, 348
Serinus canicollis, 348
Serinus hortulanus, 14
Serinus hortulanus canarius, 13
Serinus icterinus, 348
Shearwater, Dusky, 320
Shearwater, Great, 340
Shearwater, Sooty, 319
Sheldrake, Ruddy, 160
Shore-Lark, 29
Short-toed Lark, 35
Shrike, Great Gray, 55
Shrike, Lesser Gray, 53
Shrike, Pallas's Gray, 335
Siberian Ground Thrush, 335
Siberian Pectoral Sandpiper, 336
Siberian Sparrow, Brandt's, 349
Skua, Buffon's, 282
Skua, Pomatorhine, 283
Smew, 186
Snipe, Great, 278
Snipe, Jack, 280
Snipe, Red-breasted, 257
Snow Goose, Lesser, 147
Snowy Owl, 112
Sociable Lapwing, 225
Solitary Sandpiper, 336
Somateria spectabilis, 183
Somateria stelleri, 181
Sooty Shearwater, 319
Sooty Tern, 300
Soudan Crane, 357

South African Serin, 348
Sparrow, White-throated, 349
Spoonbill, 193
Spotted Cuckoo, Great, 93
Spotted Eagle, 133
Spotted Sandpiper, 246
Spur-winged Goose, 355
Squacco Heron, 206
Starling, Meadow, 348
Starling, Red-winged, 347
Starling, Rose-coloured, 4
Steller's Eider, 181
Stercorarius buffoni, 282
Stercorarius pomatorhinus, 283
Sterna anæstheta, 358
Sterna anglica, 304
Sterna caspia, 302
Sterna fuliginosa, 300
Stilt, Common, 237
Stint, American, 270
Stint, Little, 267
Stint, Temminck's, 272
Stork, Black, 197
Stork, White, 195
Strepsilas interpres, 259
Sturnella magna, 348
Surf Scoter, 176
Surnia funerea, 114
Swallow, American Tree, 351
Swallow, Red-rumped, 351
Swallow-tailed Kite, 137
Swan, American, 353
Swan, Bewick's, 145
Swan, Hooper, 144
Swan, Trumpeter, 353
Swift, Needle-tailed, 335
Swift, White-bellied, 99
Sylvia nisoria, 61
Sylvia orphea, 59

Tachycineta bicolor, 351
Tadorna casarca, 160
Tawny Pipit, 41
Teal, American, 163
Teal, Blue-winged, 164
Temminck's Stint, 272
Tengmalm's Owl, 109
Tern, Black, 308
Tern, Bridled, 358
Tern, Caspian, 302

Tern, Gull-billed, 304
Tern, Noddy, 298
Tern, Sooty, 300
Tern, Whiskered, 306
Tern, White-winged Black, 310
Thrush, Rock, 86
Thrush, Siberian Ground, 71
Thrush, White's Ground, 71
Tichodroma muraria, 47
Totanus bartrami, 245
Totanus flavipes, 250
Totanus fuscus, 252
Totanus macularius, 246
Totanus ochropus, 248
Totanus solitarius, 336
Totanus stagnatilis, 357
Tree Swallow, American, 351
Tringa acuminata, 336
Tringa acuminata pectoralis, 266
Tringa arenaria, 274
Tringa canutus, 337
Tringa fuscicollis, 261
Tringa maritima, 262
Tringa minuta, 267
Tringa platyrhyncha, 264
Tringa rufescens, 276
Tringa subarquata, 337
Tringa subminuta minutilla, 270
Tringa temmincki, 272
Trumpeter Swan, 353
Turdus iliacus, 75
Turdus migratorius, 350
Turdus pilaris, 73
Turnix sylvatica, 357
Turnstone, 259
Turtle Dove, Eastern, 331
Turtur orientalis, 331

Uria troile brunnichi, 311

Vanellus gregarius, 225
Velvet Scoter, 175
Vulture, Egyptian, 121
Vulture, Griffon, 119

Wall-Creeper, 47
Warbler, Aquatic, 67
Warbler, Barred, 61
Warbler, Great Reed, 65
Warbler, Icterine, 69
Warbler, Orphean, 59
Warbler, Rufous, 63
Waxwing, 51
Wheatear, Black-throated, 83
Wheatear, Desert, 84
Wheatear, Isabelline, 81
Whimbrel, Eskimo, 241
Whiskered Tern, 306
White-bellied Brent Goose, 156
White-bellied Swift, 99
White-billed Diver, 344
White-collared Flycatcher, 350
White Egret, Great, 209
White-eyed Pochard, 167
White-fronted Goose, 151
White-fronted Goose, Lesser, 153
White Jer-Falcon, 123
White Stork, 195
White-throated Sparrow, 349
White-winged Black Tern, 310
White-winged Crossbill, American, 6
White-winged Crossbill, European, 7
White-winged Lark, 33
White's Ground Thrush, 71
Widgeon, American, 161
Willow Wren, Yellow-browed, 57
Wilson's Petrel, 317
Wren, Ruby-crowned, 350
Wren, Yellow-browed Willow, 57

Xema sabinii, 297

Yellow-billed Cuckoo, 97
Yellow-browed Willow Wren, 57
Yellow-legged Sandpiper, 250
Yellow-rumped Seed-eater, 348

Zonotrichia albicollis, 349

www.ingramcontent.com/pod-product-compliance
Lightning Source LLC
Chambersburg PA
CBHW030407230426
43664CB00007BB/786